12/92
chicago

FIELD EXPERIMENTS AND MEASUREMENT PROGRAMS
IN GEOMORPHOLOGY

Field Experiments and Measurement Programs in Geomorphology

Edited by
OLAV SLAYMAKER
University of British Columbia, Vancouver

1991
A.A. BALKEMA / ROTTERDAM
UNIVERSITY OF BRITISH COLUMBIA PRESS / VANCOUVER

Published outside USA and Canada by
A.A. Balkema, P.O. Box 1675, 3000 BR Rotterdam, Netherlands
ISBN 90 6191 996 7

Published in USA and Canada by
University of British Columbia Press, 6344 Memorial Road, Vancouver, BC V6T 1WS, Canada
ISBN 0-7748-0351-7

Printed in the Netherlands

Table of contents

List of figures and tables

FIGURES

TABLES

Preface

This volume results from the combined efforts of corresponding members of the International Geographical Union Commission on Field Experiments in Geomorphology during the period 1976–84. The book is, therefore, in part a historical document summarising the achievements of an ephemeral IGU Commission in stimulating and recording progress in the science of geomorphology. Because experimental geomorphology barely exists in the mind of most geomorphologists the book proselytizes by attempting to demonstrate the enormous potential contribution of field experiments to geomorphology. Four task force leaders wrote individual chapters and the Commission Chairman, Olav Slaymaker, contributed a general chapter and edited the volume. Inspiration to undertake this task was provided by two earlier IGU Commission Chairmen, Professors Dr. A. Jahn and Dr. A. Rapp, and that encouragement is gratefully acknowledged here. The four task force leaders, Dr. S. Okuda of the University of Kyoto, Japan, Professor R. Bryan of the University of Toronto, Canada, Professor W. Dietrich of the University of California at Berkeley, USA (assisted by J.D. Gallinatti of the same university) and Dr. D. Walling of the University of Exeter, UK, provided outstanding leadership and co-operation which is also gratefully acknowledged. Although there are some obvious biases in the spatial coverage and representativeness of these experiments, it is thought that the typology of field experiments and the state of knowledge in the individual sub-fields is well demonstrated.

The editor is grateful to Professor Dr. H. Bremer of the University of Köln, Dr. J.A.A. Jones of University College of Wales, Aberystwyth, Professor Dr. A. Jahn of the University of Wroclaw, Poland, and Dr. M. Sweeting of Oxford University for their substantive contributions during the preparation of this volume and for helpful discussion and suggestions which have been incorporated into Chapter 1.

Introduction

OLAV SLAYMAKER
University of British Columbia, Vancouver, Canada

There is a growing concern among informed scientists that environmental deterioration is reaching potentially catastrophic dimensions. The International Council of Scientific Unions is actively promoting an International Geosphere–Biosphere Programme (frequently called Global Change), which has as its objectives 'To describe and understand the interactive physical, chemical and biological processes that regulate the total Earth System, the unique environment it provides for life, the changes that are occurring in that system and the manner in which these changes are influenced by human actions' (ICSU 1986). Geomorphology, which is the study of the form of the earth's surface, is central to the understanding of the stability of our environment. In the past, because of the assumption that changes of the earth's surface occur slowly, the study has been largely descriptive. With growth in understanding that, for example, soil erosion rates are frequently in excess of soil formation rates by a factor of ten (Fyfe et al. 1983) it is apparent that a more quantitative and experimental geomorphology is required to assess the significance of rates of change of the earth surface form. Hence the timeliness of an attempt to make a global survey of these changes and to advance a typology of experimentation in geomorphology that will allow comparative assessment of processes of earth surface change in different regions of the world.

In Chapter 1, Slaymaker discusses the varieties of field experiments in geomorphology, and the actual and potential contribution of field experiments to geomorphology. The mandate of the IGU Commission on Field Experiments (1976–84) was to promote field experimentation in geomorphology. Over 300 field experiments are referenced in this volume, so that at the quantitative level the Commission has had some success. Nevertheless, there is a sense of dissatisfaction with the rate of progress in fundamental understanding of that part of our science which is susceptible to the application of experimental method. A typology of experiments is discussed. The true experiment requires that the experimenter have control over all independent forces that are thought to influence the geomorphic effect. Such control is rarely, if ever, achievable in the field. Hence the need to discuss exploratory and less rigorous forms of experimentation, 'hybrid' and 'quasi' experiments, recognising the need to achieve a balance between rigour of experimental procedures and realistic simplification. Better and more realistic experimental designs will lead to improved progress in understanding in geomorphology.

In Chapter 2, Walling presents field experiments on particulate and solute yield at

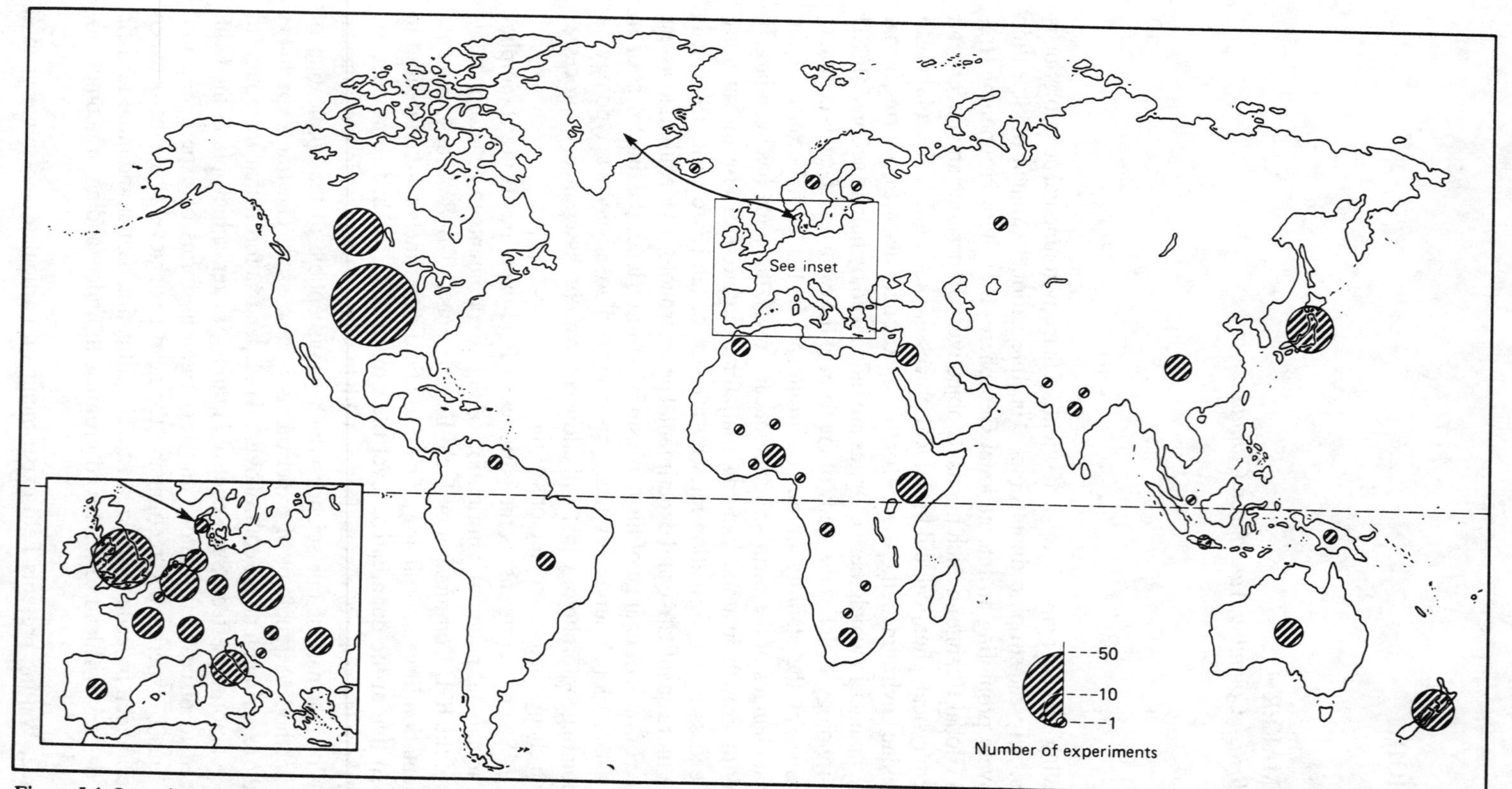

Figure I.1. Location of all geomorphic field experiments reported.

basin outlet, hydrologic processes as a basis for understanding erosion and dissolution, sediment sources and sediment budgeting, solute generation and pathways of water movement, and drainage basin modelling. In this chapter many explicit experimental designs are presented. Walling's conclusion is appropriate for the whole volume when he suggests that most present-day field experiments in geomorphology are at best quasi experiments and that these are nevertheless important because they are 'paving the way for a new generation of scientific experiments'.

In Chapter 3, Okuda discusses field experiments on slope morphometry controls, in situ-slope failures, hydrologic controls, flow phenomena, modelling and simulation, and hillslope evolution and rapid mass movement. This chapter includes some of the better examples of geomorphic field experiments, especially in the context of debris flow phenomena. There is also a strong applied emphasis. Okuda provides a summary of experience with technologically sophisticated field experiments in rapid mass movement and shows the most rapidly improving approaches to achieving control of environmental variables in the field.

In Chapter 4, Bryan discusses experiments based on runoff plots and practical land management problems, microplot and microcatchment studies to elucidate surface wash processes, laboratory experiments, and rainsplash experiments. The history of field experiments in surface wash is longer than that in other aspects of geomorphology, especially in so far as it was associated with the US Soil Conservation Service of the 1930's. Bryan identifies the history of field experimentation as far back as Wollny (1895) in West Germany. The problem which he stresses is that of integrating experimental results from a wide variety of traditions, notably soil conservation, agricultural engineering, hydraulic engineering, hydrology, forestry and geomorphology. He shows the lack of discussion between scientists working on laboratory experiments and drainage network experiments – all of whom have been exploring soil erosion at different scales. The scale problem is addressed more directly than in the other sections. The more rigorous experiments are discussed under the theme of runoff plots and practical land management problems; elsewhere, exploratory experiments are common.

In Chapter 5, Dietrich emphasizes channel mechanisms, bed topography, plan form channel geometry, channel changes and large rivers. He reports on both field experiments and laboratory and theoretical studies of significance to field experiments. 'The field

Table I.1. Number of field experiments reported by continent (1976-84).

Continent	Number of countries reporting	Number of countries with field experiments	Population (millions)	Field experiments	Ratio of Col. 5: Col. 4
Africa	5	11	252	29	0.12
Asia	5	7	2528	37	0.01
Australasia	3	3	23	26	1.13
Europe	13	15	432	123	0.28
N. America	2	2	267	81	0.30
S. America	2	2	161	5	0.03
Total	30	40	3693	301	0.08

Table I.2. Selected data for countries providing largest number of reports of field experiments relative to size and population.

	1. Population (10^6)	2. Area (10^6 km^2)	3. Reports	4. Experiments	5.	6.
Australia	16.2	7.7	8	8	0.49	1.0
Belgium	9.9	0.03	13	13	1.31	433.3
Canada	25.9	9.9	21	20	0.80	2.0
France	55.6	0.55	11	9	0.20	16.4
Great Britain	56.8	0.24	43	33	0.75	137.5
Israel	4.4	0.02	5	5	1.14	250.0
Italy	57.4	0.30	11	12	0.19	40.0
Japan	122.2	0.38	19	19	0.16	50.0
Kenya	22.4	0.58	6	9	0.27	15.5
Lesotho	1.6	0.03	1	1	0.62	33.3
Luxembourg	0.4	0.003	0	1	0	333.3
Netherlands	14.6	0.04	13	5	0.90	125.0
New Zealand	3.3	0.27	16	16	4.88	59.2
Papua New Guinea	3.6	0.46	2	2	0.56	4.4
Poland	37.8	0.31	18	18	0.48	5.8
Switzerland	6.6	0.04	3	6	0.46	150.0
USA	243.8	9.4	64	61	0.29	6.5

1. Population as indicated in Funk & Wagnalls *New Encyclopedia* 1988 Yearbook.
2. Area.
3. Reports provided by researchers from these countries.
4. Experiments located in these countries.
5. Ratio of Column 3 to Column 1.
6. Ratio of Column 4 to Column 2.

experiment is the ultimate test of any insight gained from laboratory experiments' is the underlying assumption of Dietrich. He demonstrates the exceptionally close relationship that exists in fluvial studies between laboratory and field experiments. Dietrich emphasizes the necessity for rooting such experiments in an understanding of fluid and sediment transport mechanisms.

In conclusion, scientists from thirty–four countries have co–operated in providing reports for this book (Fig. I.1); the experiments reported are being carried out in 44 countries and some indication of the representativeness of the sample is shown in Table I.1. The serious under–representation of Asian and South American countries results in part from the language barrier that accompanies an English language questionnaire. This suspicion is confirmed by Table I.2 in which it is seen that the best sampled countries include some of the Benelux countries, Poland, Switzerland, and the Commonwealth. Nevertheless, it seems safe to say that the main centres of innovation in experimental geomorphology over the past decade are included in Table I.2. It is a matter of regret that communication with our Soviet colleagues has declined in the past decade; over the next decade contributions from scientists in the People's Republic of China can be expected to foster a new centre of innovation.

REFERENCES

Bram, L. L. and N. H, Dickey (eds.) 1988. *Funk and Wagnalls New Encyclopedia 1988 Yearbook.* Funk and Wagnalls Inc., New York.

Fyfe, W.S. et al., 1983. Global tectonics and agriculture: a geochemical perspective. *Agriculture, Ecosystem and Environment* 9, 383–399.

International Council of Scientific Unions, 1986. *The international geosphere–biosphere programme: A study of global change.* Report prepared for the ICSU 21st General Assembly, Berne, 14–19 September, 1986.

CHAPTER 1

The nature of geomorphic field experiments

OLAV SLAYMAKER
University of British Columbia, Vancouver, Canada

A field experiment in geomorphology can be defined as 'a set of measurements conducted under controlled field conditions to develop and formalise some general principle about the evolution of landforms' (Slaymaker 1982). The critical part of this definition is that dealing with controlled field conditions. How controlled must the field conditions be before a field investigation can be called a field experiment? A number of explicit positions have been taken (Ahnert 1980, De Ploey and Gabriels 1980, Slaymaker et al. 1980, Slaymaker 1982, Church 1984). Of these, the most restrictive definition is provided by Ahnert (1980) and the most formally developed and closely reasoned position is that of Church (1984). There will continue to be active debate on the precise extent to which field experimental methodology can and ought to be developed in geomorphology but there can be little debate over the worthiness of the objectives to define appropriate categories of field experiment in gemorphology and to exploit various kinds of experimental approaches to the solution of geomorphic field problems.

At issue is the question of the balance between idiographic and nomothetic science, an issue which has long interested geographers (e.g. Hartshorne 1939). Geomorphology, together with other earth sciences, cannot be an exclusively nomothetic science, as Budel (1982) has explained. Many problems in geomorphology cannot be resolved by application of experimental method. But it is not a problem that geomorphology has been overwhelmed by experimental work; rather, the problem is one of underutilisation of experimental method, especially in the field. As supportive evidence for this generalisation, we note that the major English, French and German language textbooks in geomorphology do not discuss field experiments amongst their list of standard methods (e.g. Tricart 1965, Thornbury 1969, Bloom 1978, Budel 1982, Chorley et al. 1985). Balteanu (1983) provided the first regional experimental geomorphology focused on Romanian work in the Buzau Sub-Carpathians. Schumm et al. (1987) produced the first substantial monograph on experimental fluvial geomorphology. Laboratory experiments, by contrast with field experiments, are a well-recognised approach and were already central in G.K. Gilbert's geomorphology (e.g. Gilbert 1914, 1917). All authors in this volume have included reference to laboratory experiments, though the focus of their discussion and the novelty of the volume is experimentation and measurement programmes in the field.

The first objective of this chapter, then, is to define appropriate categories of geomorphic field experiments. This we believe will further the exploitation of field experimental method in geomorphology. The second objective is to demonstrate the actual and potential contribution of field experiments to geomorphological understanding.

1.1 APPROPRIATE CATEGORIES OF GEOMORPHIC FIELD EXPERIMENTS

1.1.1 *The nature of experiments in physics*

Experiments in physics may be divided into the following steps:

1. Observation of an effect;
2. Formulation of a hypothesis about an observed effect;
3. Controlled change of the independent factors thought to produce the observed effect;
4. Replication of (3) to permit falsification of hypothesis in (2);
5. Measurement of independent and dependent variables;
6. Derivation of laws governing the relations of these variables;
7. Establish range of applicability of the laws.

1.1.2 *The problem of levels of aggregation in geomorphology*

In geomorphology the observation of an effect and formulation of a hypothesis that is relevant to the evolution of landforms is a more difficult task than in classical physics.

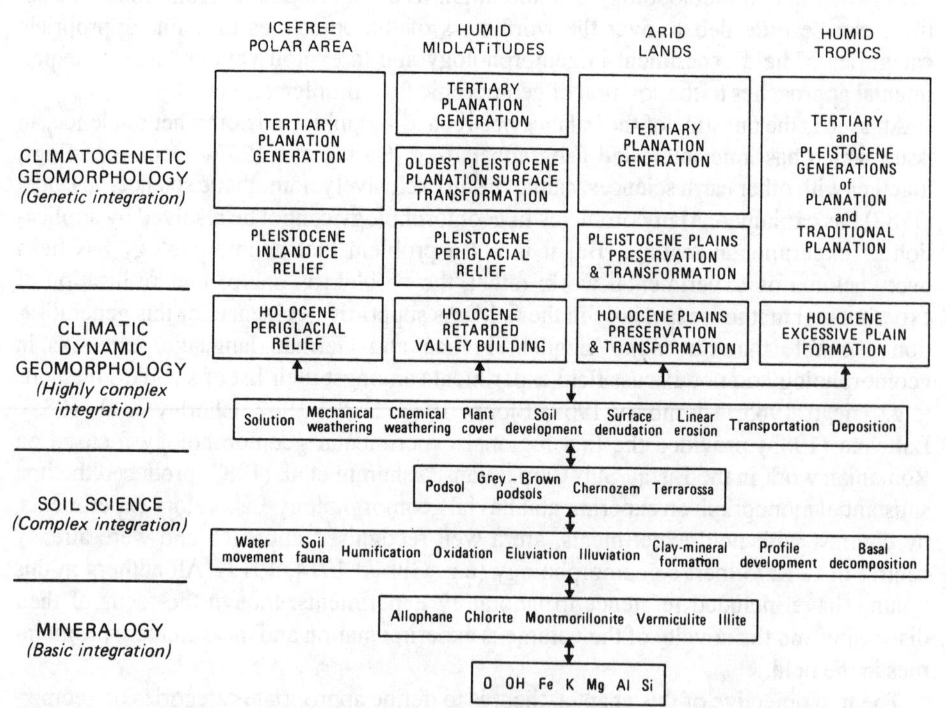

Figure 1.1. Geomorphology in the framework of the natural sciences (from Budel 1982).

This is because of the greater range of 'levels of integration' (Budel 1982) or, more accurately, levels of aggregation of the phenomena traditionally observed by geomorphologists. This is illustrated in Figure 1. 1 from the perspective of a geomorphologist who practiced climato-genetic geomorphology and came to the conclusion that at that 'level of integration' only qualitative analyses were possible.

While recognising the reality of the problem with which Budel was grappling, and acknowledging that many geomorphic problems are inaccessible through experimental method, the challenge is to define those geomorphic problems, of which there are many, which are susceptible to solution by experimental method.

1.1.3 *Varieties of field experiment in geomorphology*

Geomorphology is not alone in engaging this discussion. Quantitative ecologists (e.g. Eberhardt 1976) have recognised that field experiments in ecology cannot be equated with classical physics experiments, especially with respect to the degree of control over variables. Social scientists (e.g. Boruch and Riecken 1975), have debated the nature of experimentation in their disciplines. A discussion by Hammond (1978) on the characteristics and constraints of various modes of enquiry sets the issue in a broader context. He recognises a continuum of modes of enquiry from 'true experiment' to 'intuitive judgment'; along the continuum, the covertness of the mode of enquiry increases, active manipulation of the variables decreases and the analytical is replaced by the intuitive approach. True, hybrid, and quasi experiments, in decreasing order of variable control, are defined. In the true experiment, conditions are precisely controlled, as in a laboratory set-up; in the hybrid experiment, statistical design is exploited to ensure a high level of control; in the quasi experiment, the level of control is substantially reduced and the judgment of an experienced scientist substitutes for formal experimental design. The relationship between Hammond's typology and other suggested classifications is shown in Figure 1.2.

Church (1984) has set out the necessary and sufficient conditions for a set of geomorphological measurements to be labelled an 'experiment' as follows:

1. The measurements must be made in the context of an explicit general conceptual model of landform evolution;
2. Specific hypotheses about the evolution of landforms which are amenable to falsification must be formalised;
3. Definitions of explicit geomorphological properties and operational statements with respect to their measurement are needed;
4. A formal schedule of measurements is required;
5. A formal scheme for analysis of measurements is necessary;
6. A data collection and management system must be established.

This is a helpful checklist and emphasises the formality of scientific experiment. It does not specify the level of control required, however, and therefore is quite permissive vis-à-vis the Hammond typology discussed above.

Ahnert (1980) made the case that true experiments in geomorphology should not differ from classical experiments in physical science. His definition is so restrictive that it is questionable whether any geomorphic field experiments can be identified. Manipulation of the natural environment to the extent required by Ahnert's definition will probably ensure that effects observed are artefacts of the experimental procedure and not insights

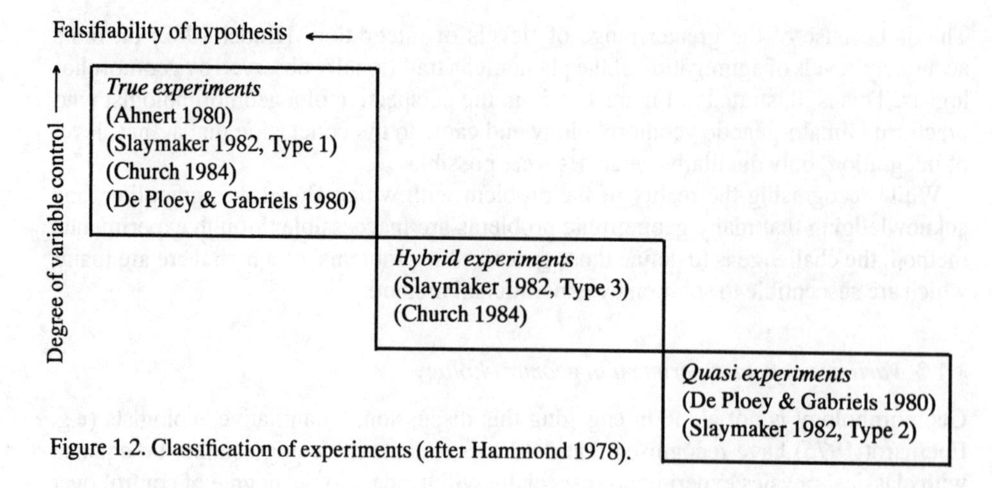

Figure 1.2. Classification of experiments (after Hammond 1978).

A. True experiments
1. Precisely controlled field conditions;
2. Mass and energy budget framework;
3. Boundary conditions of natural system are specified in space and time;
4. High level of hypothesis falsifiability.

B. Hybrid experiments
1. Spatial stratification of landscape formal sampling design required;
2. Level of variable control reduced;
3. Systems framework;
4. Hydrologic analogue is 'paired', 'side by side', or multiple watershed study.

C. Quasi experiments
1. Informed judgement used to select morphologic unit of interest;
2. Magnitude and frequency of events framework;
3. Hydrologic analogue is 'before and after' study;
4. Judgment of experienced scientist substituted for formal experimental design.

Figure 1.3. Varieties of geomorphic field experiment.

into landform evolution. De Ploey and Gabriels (1980) discussed (though not in these words) true and quasi experiments in soil loss research (pp. 66-69) and indicated the difficulty of relating these results to long-term landform evolution. Church (1984) recognised both true and hybrid experiments as valid categories of geomorphic field experiment. Slaymaker (1982) proposed three types of geomorphic field experiment, which correspond closely to the three categories suggested by Hammond (1978).

The Type 1 experiment of Slaymaker (1982) or true experiment of Hammond (1978) is rare in field geomorphology. (Fig. 1.3). The specific characteristics of such an experiment, beyond the common requirements defined by Church (1984) above, are that the field conditions must be precisely controlled so that mass and/or energy budgets can be quantitatively evaluated and the boundary conditions of the natural systems of interest can be specified in space and time. As De Ploey and Gabriels (1980) have indicated, the assumption is that the experimenter has complete control over all factors introduced in

Figure 1.4. The drained lake 'true' experiment at Illisarvik, Northwest Territories, Canada (Mackay 1981).

the experiment and that the test is repeatable by an independent experimenter. Mackay (1979) continues to provide one of the best examples of a true field experiment in geomorphology (Fig. 1.4).

The Type 3 experiment of Slaymaker (1982) or hybrid experiment of Hammond (1978) is defined as an experiment in which the landscape is stratified with the aid of formal statistical design and changes within comparable strata are monitored and analysed with respect to climatic or hydrologic forcing functions (Fig. 1.3). Because of the variability of surface geometry and soil properties, it is generally recognised that precise replication cannot be achieved in the field; moreover some acceptable level of variation in the properties of the strata should be defined beforehand. Blocking is an attempt to group experimental units on the basis of prior information so that they are substantially more homogeneous within groups than between them. The implications of this approach are that a systems framework, in which the responses of environmental systems are compared, rather than a mass and/or energy budget framework is used. The analogue in hydrologic field experiments is that of the so-called paired watershed study. The nature of the statistical control in such experiments is discussed by Church (1984, pp. 568-569). Bovis (1978) provides an instructive geomorphic example, and Slaymaker (1972) is an early example of this kind of experiment (Fig. 1.5).

The Type 2 experiment of Slaymaker (1982) or quasi experiment of Hammond (1978) is the field experiment which, while it satisfies each of the six criteria stated by Church (1984), has the lowest level of control over environmental variables. Church, in relation to Type 2 experiments, suggested that 'there is no effective control over the subject of

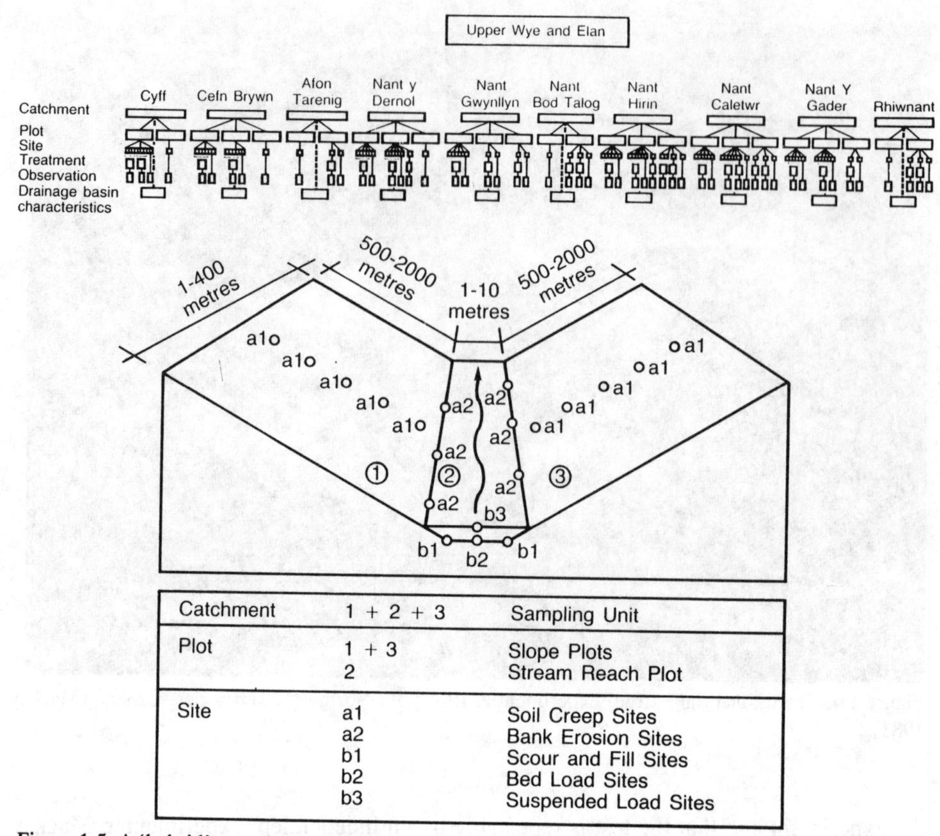

Figure 1.5. A 'hybrid' experiment from central Wales (Slaymaker 1972).

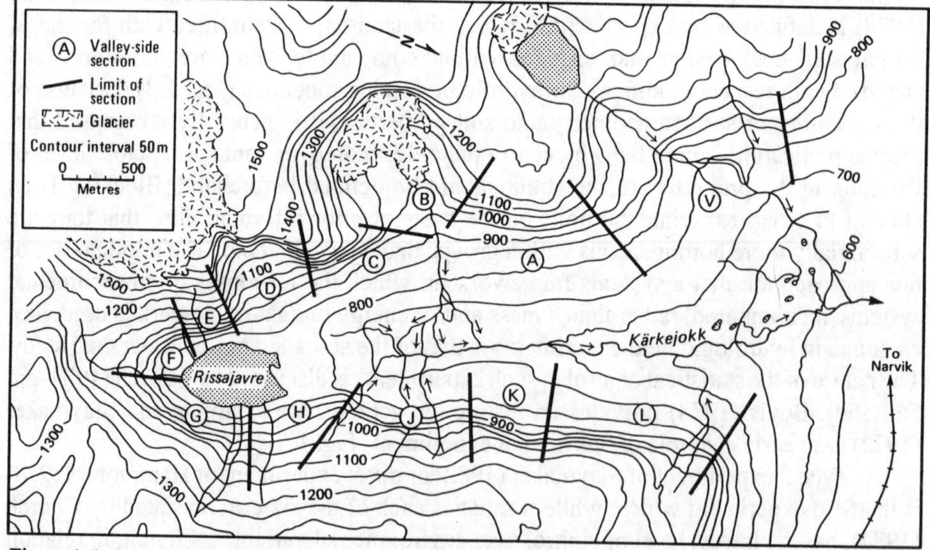

Figure 1.6. A 'quasi' experiment conducted at Kärkevagge, northern Sweden (Rapp 1960).

interest' (pp. 571 and 577). The control which is present in such quasi experiments is defined by the informed judgment of the gemorphologist in selecting monitoring sites within the spatial limits of a specific landform. The analogue in hydrologic field experiments is that of the so-called 'before and after' study of watershed response to land use change. In the latter case, the control unit is the watershed, defined by the informed judgment of the hydrologist as the environmental system of interest. (Fig. 1.3). Rapp (1960) provides one of the classic examples of this kind of experiment (Fig. 1.6).

Although discussion on the nature of geomorphic field experiments has been engaged only recently, a century of field experimentation by geomorphologists has passed (Davison, 1888). Hudson (1957), Schumm and Khan (1972), Pissart (1973) and Moeyersons (1975) provide excellent examples of successful field experimentation in geomorphology and the study of beach form in coastal geomorphology has long attracted experimentalists (King 1972, Komar 1976).

With respect to the nature of geomorphic field experiments it seems important at this stage in the development of our subject to be as inclusive as possible, at the same time recognising the need to improve the sophistication of our experimental methods and, where possible, to move from quasi-experiments to hybrid experiments. In this volume authors have followed a strategy of inclusiveness. This is a deliberate decision so that the extent of the challenge can be evaluated. A number of the quasi experiments considered are very little more than measurement programmes but they are included in so far as a specific hypothesis is being tested and/or novel approaches to design and instrumentation are included.

1.2 THE ACTUAL AND POTENTIAL CONTRIBUTION OF FIELD EXPERIMENTS IN GEOMORPHOLOGY

The actual contribution of research using field experiments in geomorphology is already substantial in spite of the fact that experimental geomorphology barely exists in the mind of many geomorphologists. Wurm (1935) investigated erosional development of slopes; Lewis (1944) predicted the complex response of river systems from experimental work; Ackers (1964) demonstrated the similarity of flume and river geometries; Johnson (1970) made explicit a number of problems relating to scale modelling; Gavrilovic (1972) discussed river profile development under varied climate inputs and Petts (1979) showed the value of 'inadvertent' experiments (Church 1984). In this book the major contributions of field experiments to our understanding of landscape change are:

1. An improvement in conceptualisation of the mode of operation of contemporary geomorphic processes has occurred (see Sections 2.7, 3.5, 4.3 and 5.2). Field experiments have contributed directly because of the requirement that experimental hypotheses about landform development be couched in falsifiable form.

2. There is rapidly improving understanding of the absolute rates of operation of geomorphic processes and their geographical variations (see Sections 2.6, 3.7, 4.3 and 5.5) and

3. The elusive and long sought after links between slope and river channel processes are being unravelled. Methodologically, this is well illustrated by the nearly universal adoption of a version of sediment budgetting or mass balance framework for slopes, stream reaches and whole basins (see Sections 2.4, 2.5, 2.6, 3.7, 4.6, 5.2, 5.5, and 5.6).

As far as potential contribution is concerned, 'what is required is a programme of planned experiments designed to generate the critical data needed for model evaluation' (Burt and Walling 1984, p. 14). For too long many geomorphologists, whether experimental or not, have conducted independent enquiries which have not proved to be additive. It is thought that experimental work carried out by teams of researchers is more likely to prove additive (Church et al. 1985).

Further standardisation of field experimental procedures is urgently needed in geomorphology as this leads to the possibility of comparing results from place to place. A recent example of this problem is provided by a paper and exchange of letters in Earth Surface Processes and Landforms (Saunders and Young 1983, Kotarba and Starkel 1985, Saunders and Young 1985).

New techniques and instrumentation networks have been developed (see Sections 2.4, 2.5, 3.5, 4.2 and 5.2). Until field experiments were devised it was not apparent that there was a need for such a variety of devices that are both ingeniously simple and technologically sophisticated. Expanded global cooperation on the basis of standardised experimental procedures will produce readily comparable data (cf. Sections 2.6, 3.1, 4.2, 4.3, 5.3, 5.4 and 5.5).

The development of field experiments in geomorphology has established links with other environmental sciences and the prediction of environmental impact of human activities with respect to landform change has become possible (see Chapter 5 for some increasingly successful illustrations as well as Sections 2.6, 3.3 and 4.2). Renewed effort is called for so that predictive statements about landform response can be developed.

As pointed out by Burt and Walling (1984) studies of contemporary geomorphic process provide the major focus of attention in geomorphology at present. There are at least three reasons for this interest:

1. Our inadequate understanding of contemporary geomorphic processes as land forming agents has been overlooked until recently.

2. Global environmental change requires that geomorphologists consider future scenarios on the basis of contemporary process understanding, and

3. Society's modification of the earth's surface requires improved predictive statements about geomorphic response.

It is worth repeating that 'field experimentation will be of increasing importance in maintaining links with the landscape and the magnitude and complexity of its processes and resulting landform' (Slaymaker et al. 1978). This volume demonstrates unequivocally a range of new questions about landscape that field experimentation has provoked. Most of these new questions require answers before satisfactory models of relief development can be provided. At the same time, there are pressing reasons for geomorphologists to direct increasing attention to field work that will provide additive data such that prediction of lithospheric aspects of global change can be improved.

REFERENCES

Ackers, P., 1964. Experiments on small streams in alluvium. *J. Hydraulics Div., Amer. Soc. Civ. Eng.* 90, 1-37.

Ahnert, F., 1980. A note on measurements and experiments in geomorphology. *Z. Geomorph. Supp. Bd.* 35, 1-10.

Balteanu, D., 1983. *Experimentul de teren in geomorfologie.* Acad. Romanian Socialist Republic, Bucharest.

Bloom, A.L., 1978. *Geomorphology*, Prentice-Hall.

Boruch, R.F. and H.W. Riecken, 1975. *Experimental testing of public policy*, Westview.

Bovis, M.J., 1978. Soil loss in the Colorado Front Range.*Z. Geomorph. Supp. Bd.* 29, 10-21.

Budel, J., 1982. *Climatic geomorphology* (Trans L. Fischer and D. Busche), Princeton.

Burt, T.P. and D.E. Walling (eds.), 1984. *Catchment experiments in fluvial geomorphology*. Geobooks, Norwich, UK.

Chorley, R.J., S.A. Schumm and D.E. Sugden, 1984. *Geomorphology*. Methuen.

Church, M., 1984. On experimental method in geomorphology. In T.P. Burt and D.E. Walling (eds.), *Catchment experiments in fluvial geomorphology*, 563-580.

Church, M., B. Gomez, E.J. Hickin and O. Slaymaker, 1985. Geomorphological sociology. *Earth Surface Processes and Landforms*, 10, 539-540.

Davison, C., 1888. Note on the movement of scree material. *Q.J. Geol. Soc. London*, 44, 232-238.

De Ploey, J. and D. Gabriels, 1980. Measuring soil loss and experimental studies. In M.J. Kirkby and R.C.C. Morgan (eds.), *Soil erosion*, Wiley, 63-108.

Eberhardt, L.L., 1976. Quantitative ecology and impact assessment. *J. Environ. Management*, 4, 27-70.

Gavrilovic, P., 1972. Experimente zur klimageomorphologie. *Z. Geomorph.* 16, 315-331.

Gilbert, G.K., 1914. The transportation of debris by running water. *U.S. Geol. Surv. Prof. Paper* 86.

Gilbert, G.K., 1917. Hydraulic mining debris in the Sierra Nevada.*U.S. Geol. Surv. Prof. Paper* 105.

Hammond, K.R., 1978. Toward increasing competence of thought in public policy formation. In K.R. Hammond (ed.), *Judgment and decision in public policy formation*, Westview, 11-32.

Hartshorne, R., 1939. The nature of geography. *Ann. Assoc. Amer. Geog.*, 29.

Hudson, N.W., 1957. The design of field experiments on soil erosion. *J. Agric. Engineering Research*, 2, 56-65.

Johnson, A.M., 1970. *Physical processes in geology.* Freeman.

King, C.A.M., 1972. *Beaches and coasts.* Edward Arnold.

Komar, P.D., 1976. *Beach processes and sedimentation*. Prentice-Hall.

Kotarba, A. and L. Starkel, 1985. Rates of surface processes on slopes, slope retreat and denudation: a comment. *Earth Surface Processes and Landforms*, 10, 83-84.

Lewis, W.V., 1944. Stream trough experiments and terrace formation. *Geol. Mag.* 81, 241-253.

Mackay, J.R., 1981. An experiment in lake drainage, Richards Island, N.W.T.: a progress report. *Geol. Surv. Canada Paper* 81-1A, 63-68.

Moeyersons, J., 1975. An experimental study of pluvial processes on granite gruss. *Catena*, 2, 289-308.

Petts, G.E., 1979. Complex response of river channel morphology subsequent to reservoir construction. *Prog. Phys. Geog.* 3, 329-362.

Pissart, A., 1973. Résultats d'éxperiences sur l'action du gel dans le sol. *Biul. Peryglacjalny*, 23, 101-113.

Rapp, A., 1960. Recent development of mountain slopes in Karkevagge and surroundings. *Geografiska Annaler* 42A, 73-200.

Saunders, I. and A. Young, 1983. Rates of surface processes on slopes, slope retreat and denudation. *Earth Surface Processes and Landforms*, 8, 473-501.

Saunders, I. and A. Young, 1985. Rates of surface processes on slopes, slope retreat and denudation: a reply. *Earth Surface Processes and Landforms*, 10, 85-86.

Schumm, S.A., 1985. Explanation and extrapolation in geomorphology in seven reasons for geologic uncertainty. *Trans. Jap. Geomorph. Union* 6, 1-18.

Schumm, S.A. and H.R. Khan, 1972. Experimental study of channel patterns. *Geol. Soc. Amer. Bull.* 83, 1755-1770.

Schumm, S.A., M.P. Mosley and W.E. Weaver, 1987. *Experimental fluvial geomorphology*. Wiley.

Slaymaker, O., 1972. Patterns of subaerial erosion in mid-Wales. *Trans. Inst. Brit. Geog.* 55, 47-68.

Slaymaker, O., 1982. The nature of field experiments in geomorphology. *Studia Geomorph. Carpatho-Balcanica*, 15, 11-17.

Slaymaker, O., T. Dunne and A. Rapp, 1980. Preface to Geomorphic Experiments on Hillslopes, *Z. Geomorph. Supp. Bd.* 35, v-vii.

Slaymaker, O., A. Rapp and T. Dunne, 1978. Preface to Field Instrumentation and Geomorphological Problems, *Z. Geomorph. Supp. Bd.* 29, v-vii.

Thornbury, W.D., 1969. *Principles of geomorphology*. Wiley.

Tricart, J., 1965. *Principes et méthodes de géomorphologie*. Masson et Cie, Paris.

CHAPTER 2

Drainage basin studies

D. E. WALLING
University of Exeter, UK

2.1 THE BASIS

Several discussions of the precise definition of geomorphological 'experiments' and the extent to which particular field measurement programmes merit such designation have appeared in the recent literature (e.g. Ahnert, 1980; Slaymaker, 1980; Slaymaker et al., 1980; Church, 1984). It is not the purpose of this report to add to this discussion, although it is perhaps pertinent to refer to the detailed discussion presented by Church (1984) and to his conclusion that there were few current field investigations that came near to fulfilling all the criteria that should be met by a 'scientific experiment'. He indicates that most existing studies should be classified as 'case studies' rather than true 'experiments' but he emphasises that such case studies: 'are an essential part of the complete enquiry that develops conceptual understanding.'

If, as Church (1984) argues, true scientific experiments can only stem from a well-developed and clearly-defined conceptual base, then such case studies should be seen as paving the way for a new generation of more scientific experiments. Whereas the basic precepts of many scientific disciplines were established a century or more ago, geomorphology is a relatively young science. It should, for example, not be forgotten that the existence and potential significance of such drainage basin processes as throughflow was fully recognized only 15 years ago (e.g. Kirkby and Chorley, 1967) and that our understanding is still developing. No attempt will therefore be made in this report to take a rigid view of the term 'experiment', to classify the studies reviewed in terms of the extent to which they merit this designation, or to exclude work on the basis of such definitions. Rather, it will attempt to look at the nature of the drainage basin studies that are actually being undertaken in different areas of the world and at their objectives, methods, and achievements.

2.2 THE SURVEY

In attempting to assemble material for this review, use has been made of two major sources. First, the collection of papers presented at the paper sessions of the UK meeting of the Commission in August 1981 at Exeter and Huddersfield (Burt and Walling, 1984),

Table 2.1. Summary of drainage basin studies included in this review.

No.	Country	Researchers	Catchment and location	Area	Period of study	Major objectives* 1	2	3	4	5	Sample reference
1	Canada	T.M.Gallie & O.Slaymaker	Goat Meadows Watershed, Coast Mountains, British Columbia	2 ha	1979-	x			x		Gallie & Slaymaker (1984)
2	Canada	C.C.Smart	Mt. Castleguard, Rocky Mountains, Alberta (Glacierized karst)	c. 45 km^2	1979-80		x				Smart & Ford (1982)
3	Canada	R.B.Bryan & W.K.Hodges	Microcatchments, Dinosaur Prov. Park, Alberta	10.6-44.3 m^2	1980-	x	x	x			Bryan & Hodges (1984)
4	Canada	I.A.Campbell & R.B.Bryan	Dinosaur Park Catchment, Dinosaur Prov. Park, Alberta	0.377 km^2	1981-	x	x	x	x		Bryan & Campbell (1982)
5	USA	F.J.Swanson & co-workers	Small watersheds in H.J.Andrews Experiment Forest, Oregon	9-101 ha	1950s-	x		x			Swanson et al. (1982)
6	USA	R.M.Rice & co-workers	Casper Creek, N.California	5.1 km^2 and 12 sub-catchments	1983-	x		x			Details from Pacific Southwest Forest and Range Experimental Station, Arcata, Calif., USA
7	USA	A.K.Lehre	Lone Tree Creek drainage basin, California	1.74 km^2	1971-74	x		x			Lehre (1982)
8	USA	E.D.Andrews	Piceance Creek, Colorado	1632 km^2 & sub-basins		x					Andrews (1983)
9	USA	USDA Sedimentation Laboratory	Goodwin Creek Research Catchment, North-central Mississippi	21.4 km^2 sub-basin	1977	x		x		x	Decoursey (1982a)
10	USA	A.A.Afifi & O.P.Bricker	Mill Run Basin Virginia	3.2 km^2	1982-	x		x			Afifi & Bricker (1983)
11	USA	N.M.Johnson, G.E.Likens & co-workers	Hubbard Brook catchments, New Hampshire	12-43 ha	1955	x			x	x	Likens et al. (1977)
12	Brazil	A.L.C.Netto	Cachoeira basin Tijuca National Park, Rio de Janeiro	3.5 km^2	1976-	x	x	x			Contact Professor A.L.C.Netto, Institute of Geoscience, Federal Univ. of Rio de Janeiro

13	UK	S. Nortcliff & J. B. Thornes	Barro Branco basin Nr. Manaus Amazonia	1.5 km^2 & instrumented plot	1977-		x		x		Nortcliff & Thornes (1981)
14	UK	J. M. Reid & co-workers	Glendye catchment, Grampian Highlands	41.2 km^2	1977-	x			x		Trudgill et al. (1984)
15	UK	S. Trudgill & co-workers	Whitwell Wood	c. 4 km^4	1978-	x	x		x		Trudgill et al. (1984)
16	UK	T. P. Burt & co-workers	Shiny Brook basins	0.09 & 0.11 km^2	1978-	x	x	x	x		Burt & Gardiner (1984)
17	UK	M. D. Newson, G. Leeks & co-workers	Plynlimon catchments, mid Wales	8.7 & 10.55 km^2 plus sub-basins		x	x	x	x	x	Newson (1980), Roberts et al. (1983)
18	UK	J. A. A. Jones	Maesnant catchment, mid Wales	0.54 km^2	1979-		x	x			Jones & Crane (1984)
19	UK	I. D. L. Foster & co-workers	Merevale basin (Lake catchment) N. Warwickshire	1.95 km^2	1980-	x		x	x		Foster et al. (1983)
20	UK	M. G. Anderson & P. E. Kneale	Winsford catchment nr Bristol	0.8 km^2	1978-81		x			x	Anderson & Keale (1982)
21	UK	T. P. Burt & M. G. Anderson	Bicknoller catchment, Quantock Hills	0.6 km^2	1975-	x	x		x	x	Anderson & Burt (1978)
22	UK	K. J. Gregory & A. M. Gurnell	Highland Water catchment and sub-basins, New Forest	11.4 km^2	1974-	x	x	x	x		Gurnell & Gregory (1984)
23	UK	D. E. Walling, B. W. Webb & co-workers	Jackmoor Brook & Dart catchments, Devon	9.8 & 46 km^2	1975-	x		x	x		Peart & Walling (1982)
24	UK	D. E. Walling, B. W. Webb & co-workers	Exe basin, Devon	c. 1500 km^2	1971-	x		x	x		Webb & Walling (1983)
25	UK	A. G. Williams, J. L. Ternan & co-workers	Narrator basin, Devon	4.75 km^2	1977	x	x	x	x		Williams et al. (1983)
26	UK	A. Thomas, T. P. Burt & co-workers	Slapton Wood catchment, S. Devon	0.94 km^2	1969	x	x	x	x		Burt et al. (1983)
27	UK	D. N. Collins	Gornera Glacier Catchment, Valais	82 km^2 (83% glacierized)	1975-	x		x	x		Collins (1983)

Table 2.1 (continued).

No.	Country	Researchers	Catchment and location	Area	Period of study	Major objectives* 1	2	3	4	5	Sample reference
28	UK	A. M. Gurnell & co-workers	Tsidjiore Nouve Glacier catchment, Valais	4.8 km^2	1977-	x		x	x		Gurnell (1982)
29	Denmark	B. Hasholt	Mitdluagkat basin, E. Greenland	8 km^2	1972-	x					Hasholt (1976)
30	Sweden	B. Calles, U. M. Andersson-Calles & co-workers	Velen, Kassjoan & Lapptrasket Representative Basins	45, 165, 1004 km^2	1969-	x					Andersson, Calles & Eriksson (1979)
31	Finland	M. Tikkanen, M. Seppala & O. Heikkinen	2 basins, Paajarvi Region, Lammi	6 & 8 km^2	1982-	x					Contact Dr. O. Heikkinen, Dept. of Geography, University of Helsinki, Finland
32	Poland	W. Froehlich	Homerka basin, Beskidy Mts.	19.6 km^2	1971-	x	x	x	x		Froehlich (1983)
33	Poland	L. Kaszowski & K. Krzemien	Chocholowski & Starorobocianski basins, W Tatra Mts.	38.8 & 8.8 km^2	1976-	x		x			Krzemien (1982)
34	Poland	J. Szymanski, W. Rojek & L. Tymrakiewicz	Mielnica basin Trzebnica Mts. & Bogoryja basin Sudety Mts.	6.7 & 4.5 km^2	1976-	x		x			Contact Professor J. Szymanski, Institute of Land and Forest Reclamation, Agricultural College of Wroclaw, Poland
35	Poland	M. Madejski	6 tributary basins of Raba River, Carpathian Mts.	32-77 km^2	1976-	x					Madejski (1982)
36	Poland	H. Gladki & co-workers	Targaniczanka basin	13.8 km^2	1978-			x			Gladki & Czaszynski (1981)
37	Poland	B. Janiec	Upper Sanna catchment, Lublin uplands	76 km^2	1981	x					Contact Dr. B. Janiec, Dept. of Hydrography, Marie Curie Sklodowska University, Lublin
38	Poland	C. Rzepa, T. Ciupa & T. Biernat	Czarna Nida, Biala Nida & Lozosina basins, Central Poland	758, 1032 & 312 km^2	1977-	x					Contact Dr. C. Rzepa, Institute of Geography, Pedagogical College, Kielce, Poland
39	Poland	H. Maruszczak & co-workers	Bystra & Uherka basins	285 & 429 km^2	1973-	x					Contact Prof. H. Maruszczak, Dept. of Physical Geography, Marie Curie Sklodowska University, Lublin, Poland

40	Poland	C. Pietrucien & co-workers	Tazyna, Struga Zielona & Struga Torunska basins, Polish Lowlands	534, 355 & 370 km^2	1973-	x					Contact Dr. C. Pietrucien, Institute of Geography, Mikolaj Kopernik Univ. Torun, Poland
41	Poland	E. Gil & A. Weic	Bystrzanka & Bielanka Catchments, Beskid Niski Mts.	13.6 & 12.6 km^2	1969-	x	x	x	x		Welc (1978)
42	Netherlands	P. D. Jungerius & A. Imeson & co-workers	Several small catchments in the Keuper region of Luxembourg near Ermsdorf	c. 1 km^2	1975-	x	x	x	x		Imeson & Von Zon (1979)
43	Netherlands	J. Roels	Small basin, Ardeche	8250 m^2	1977-	x		x			Contact Dr. J. Roels, Geografisch Institut, Rijksuniversiteit, Utrecht, Netherlands
44	Belgium	P. Buldgen & J. Remacle	2 small catchments Haute Ardenne	0.8 km^2	–	x	x		x		Contact P. Buldgen, Dept. of Botany, University of Liege, Belgium
45	Belgium	A. Pissart & co-workers	River Meuse at Liege & River Ourthe	–	–	x					Contact A. Pissart, Dept. of Geomorphology, University of Liege, Belgium
46	France	F. Lelong & co-workers	Valat de la Sapine, Valat de la Latte, Valat des Cloutasses, Mont Lozere	0.54, 0.20 & 0.75 km^2	1981-	x		x			Dupraz et al. (1982)
47	France	B. Ambroise & co-workers	Ringelbach, Petite Fecht & Fecht basins, Vosges Mts.	0.36, 12.3 & 450 km^2	1975-		x			x	Association Géographique d'Alsace (1982)
48	France	J. Tricart	Baschney basin, N. Vosges	2.7 km^2	1980-		x				Contact Prof. J. Tricart, Centre de Géographie Appliquée, Université Louis Pasteur, Strasbourg
49	France	C. Martin	Rimbaud basin, Massif des Maures	1.46 km^2	1976-	x			x		Martin (1981)
50	Spain	M. J. Liedo Solbes & co-workers	Barranc de L'Avic basin, nr Tarragona	0.55 km^2	1981-	x		x	x		Contact M. J. Liedo, Solbes Dept. of Biology, Universidad de Alicante
51	Spain	M. Sala	Fuirosos basin and sub-basins	16.6 km^2	1977-	x		x			Sala (1981)
52	Switzerland	H. M. Keller	Alptal catchments (6)	0.52-1.55 km^2	1967-	x			x		Keller & Strobel (1982)
53	Italy	P. Tacconi & co-workers	Virginio and Pesciola catchments & sub-basins	62 km^2	1977-	x					Becchi et al. (1979)
54	Italy	G. S. Tazioli & co-workers	Two basins in Basillicata & Calabria	63.5 & 4.7 km^2	1976-	x					Tazioli (1981)

Table 2.1 (continued).

No.	Country	Researchers	Catchment and location	Area	Period of study	Major objectives* 1	2	3	4	5	Sample reference
55	Romania	D. Balteanu & co-workers	Drghici, Tatar & Porcareata basins, S. E. Carpathians	4.87, 0.75 & 1.3 km^2	1960			x			Balteanu et al. (1984)
56	Israel	A. Yair & co-workers	Sede Boger, N. Negev	1.13 ha	1973-		x	x	x		Yair (1981)
57	Israel	A. P. Schick & co-workers	Nahal Yael basin, S. Negev	0.5 km^2	1965-	x		x			Lekach & Schick (1982)
58	Nigeria	L. K. Jeje & co-workers	Small catchments near Ife, Nigeria	6 2nd, 15 3rd order	1981-	x					Jeje & Nebega (1983)
59	Zaire	M. Lootens	Mikuta basin and Kafabu basin	4.5 & 1540 km^2	1979-	x					Contact Prof. M. Lootens, Dept. of Geographie, Univ. Nationale du Zaire, Lubumbashi
60	Zimbabwe	J. P. Watson	Juliasdale and Rusape basins	0.91 & 7.33 km^2	1971-	x					Owens & Watson (1979)
61	Lesotho	Q. K. Chakela	Roma Valley and Maliele catchments (9)	c. 1-10 km^2	1973-	x		x			Chakela (1981)
62	South Africa	J. S. le Roux & Z. N. Roos	Bulbergfontein & Wuras Dam Catchment	4.77 & 666 km^2	1978	x					Le Roux & Roos (1979, 1982)
63	China	Mou Jinze & Meng Qingmei	Tuanshangou catchment, Middle Yellow River basin	0.18 km^2	1965-	x		x			Mou & Quingmei (1981)
64	China	Tang Bangxing & co-workers	Jianjia gully, Yunnan	47.1 km^2	1965-	x					Kang & Zhang (1984)
65	Japan	T. Tanaka, S. Takayama & co-workers	Hachioji basin Japan	2.2 ha	1980-	x	x	x			Tanaka (1982)
66	Indonesia	L. A. Bruijnzeel	Kali Mondo basin, south-central Java	19 ha	1976-78	x	x		x		Bruijnzeel (1983)
67	Australia	M. Bonell & co-workers	North and South Creek basins, N. Queensland	18.3 ha, 25.7 ha	1975-	x	x				Bonell et al. (1982)
68	Australia	R. J. Loughran & co-workers	Maluna Creek basin Hunter Valley	1.7 km^2	1978-	x		x			Loughran et al. (1982)

69	Australia	W.A. Rieger & L.J. Olive	Karuah Catchments Hunter Valley, NSW (7)	11-65 ha	1981-	x				Contact W.A. Rieger, Dept. of Geography, Royal Military College, Duntroon, ACT, 2600
70	Australia	L.J. Olive & W.A. Rieger	Eden catchments Southeast NSW (6)	72-212 ha	1975-	x				Rieger et al. (1982)
71	Australia	B. Finlayson	Myrtle Creek, No. 1 catchment	25 ha	1980-	x				Contact Dr. B. Finlayson, Dept. of Geography, Univ. of Melbourne, Parksville, Victoria
72	New Zealand	C. L. O'Loughlin, A. J. Pearce, L. K. Rowe & co-workers (New Zealand Forest Service)	Several groups of catchments at		1974-					O'Loughlin et al. (1978, 1980, 1982)
			Maimai (9)	1.6-8.3 ha		x		x	x	
			Big Bush (4)	4.8-20 ha		x		x	x	
			Tairua (2)	3.2 & 20 ha		x		x	x	
			Ashley (2)	15 & 23 ha		x		x	x	
			Glendhu (2)	218 & 310 ha		x		x	x	
73	New Zealand	G.A. Griffiths & M.J. McSavaney	Upper Cropp basin, S. Island	12.2 km^2	1980-	x				Contact Dr. G.A. Griffiths, Water and Soil Division, Ministry of Works & Development, Christchurch
74	New Zealand	G.A. Griffiths & D.A. Hicks	Dry Acheron Basin, S. Island	21 km^2	1980-	x		x		Griffiths & Hicks (1980)
75	New Zealand	J.A. Hayward	Torlesse catchment	3.85 km^2	1973-	x		x		Hayward (1980)

*See p. 26 for explanation of objectives.

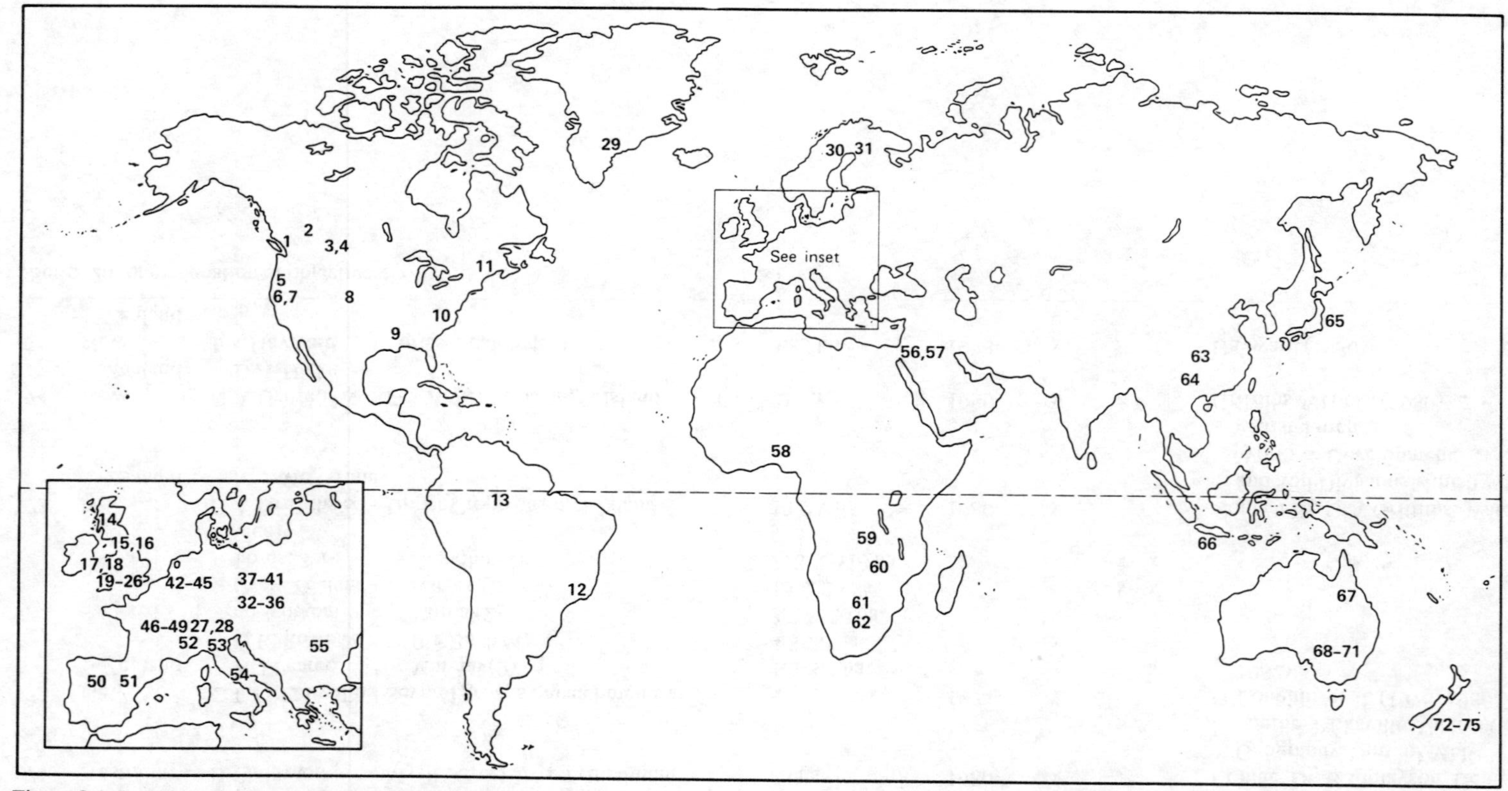

Figure 2.1. Location of drainage basin experiments reported.

provided a valuable sample of current drainage basin studies in many areas of the world, including New Zealand, Brazil, the People's Republic of China, as well as Europe and North America. Secondly, a plea for information on current drainage basin investigations was sent to over 70 corresponding members of the Commission known to be actively involved in such work, along with a request that it should be forwarded to other relevant workers in their country. The response to this exercise was somewhat disappointing but over 30 responses were received. To these sources has been added personal knowledge of a number of other catchment studies with which the writer has come into contact in recent years (Table 2.1).

Information on a total of 75 drainage basin studies which are currently in progress or have been recently concluded was collated. The global distribution of these studies is shown on Figure 2.1. It is accepted that more investigations could have been included had additional efforts been made to track down their details, but it is suggested that this sample provides a representative basis for reviewing the current status of work in this field. Further details concerning each study and relating to location, workers involved, basin size, major objectives and available publications is provided in Table 2.1. From this it may be noted that not all studies are operated specifically by geomorphologists, because an attempt has been made to embrace studies undertaken by workers in other disciplines which nevertheless include a geomorphological perspective or component. The review of these 75 studies will consider their organization and objectives, the methods and instrumentation employed and the results obtained. Inevitably this review will involve subjective interpretation, particularly where only limited documentation was available, and the writer would like to apologize for any misrepresentation that might occur in the following discussion.

2.3 ORGANIZATION AND OBJECTIVES

Considering the range of studies listed in Table 2.1, an important distinction may be drawn between those where the investigation was established specifically for the geomorphological objective listed, and those where the major justification for the establishment of the instrumented catchment must be sought elsewhere. Instrumented catchments have been set up for a wide range of hydrological research (e.g. Rodda, 1976) and, in view of the considerable costs that may be involved in instrumentation, there are clear advantages in geomorphologists making use of these pre-existing facilities. In some instances, the geomorphological investigation is an integral part of the wider programme conceived by the operating agency, but again benefits can accrue from the larger budget available for instrumentation. In most cases, drainage basins used solely for geomorphological experiments have been instrumented with limited financial resources and in consequence the basic flow gauging installations are generally less elaborate than those encountered in hydrological studies.

The work of Swanson and his colleagues (Study 5) in the H.J. Andrews Experimental Watersheds in Oregon provides an excellent example of where the organization involved, in this case the US Forest Service, has included geomorphological investigations within the scope of its research and where these have been able to make use of the existing basic instrumentation. A similar situation exists with the work of Johnson and Likens and their co-workers in the Hubbard Brook Experimental Forest (Study 11), of Newson and his colleagues within the Plynlimon catchments established by the UK Institute of Hydrology (Study 17) and of Keller in the Alptal basins operated by the Swiss Forest Service (Study 52). Examples of where workers from other organizations have profited by operating within an existing catchment established for hydrological research are provided by Studies 30, 69, 70 and 71. Reference must also be made to a number of other situations where the location of the study basin has been governed by the existence of a streamflow gauge operated by a statutory monitoring authority, although these studies generally involve larger river basins (e.g. Studies 14, 40 and 45). In the majority of cases, however, the detailed requirements of particular studies have resulted in the selection of a catchment specifically for the purpose in hand, although detailed analysis would doubtless reveal that a variety of subsidiary factors, such as the availability of access roads or the existence of a cooperative landowner, conditioned the final choice.

At a more fundamental level, the studies listed in Table 2.1 embrace a wide variety of objectives. However, it is suggested in the rubric that these can essentially be reduced to five major themes, with some studies including more than one. In this context a useful distinction can be made between those studies that are primarily concerned with monitoring the material output from a drainage basin (Theme 1) and those that also investigate the processes operating within the catchment (Themes 2-5). These five major themes may be listed as follows:

1. Measurement of particulate and solute yield at the catchment outlet as a means of documenting rates of material transport and of denudation.

2. Elucidation of runoff and other basic hydrological processes operating within the basin with the contention that such work constitutes an essential prerequisite for understanding processes of erosion and dissolution. Polish investigators have appropriately referred to such work as investigation of the processes of 'water circulation'.

3. Elucidation of the processes involved in detachment and transport of particulates, evaluation of the relative importance of specific sediment sources and establishment of sediment budgets.

4. Elucidation of the processes involved in solute generation and dissolution and more particularly the relationship of solute mobilization to rock weathering mechanisms and pathways of water movement and of the spatial distribution of dissolution rates.

5. Assembly of detailed data on catchment behaviour (i.e. runoff, sediment and solutes) in order to test and develop mathematical models.

In virtually all cases the emphasis is on documenting and understanding the operation of contemporary processes of landscape development. Few if any studies explicitly incorporate attempts to relate these to questions of longer-term relief development, although this doubtless represents an implicit and ultimate objective of many studies.

The objectives of individual studies vary considerably in complexity. Thus, for example, in some cases the objective is essentially to determine the magnitude of suspended sediment and dissolved load output from a drainage basin and perhaps to compare the dissolved and particulate components, whereas in others this may be coupled with a more demanding assessment of the sediment budget of the drainage basin or the spatial distribution of chemical denudation. To some extent the degree of complexity reflects the financial resources available to establish and operate a study, but it is also possible to detect a relationship between the complexity of the questions posed and the existing level of knowledge in a particular country. Thus if little or nothing is known about rates of landscape development in an area, basic measurements of sediment and solute output from a drainage basin may prove invaluable in providing an initial assessment of denudation rates. However, once these have been established, there will doubtless be a desire to expand the conceptual base by undertaking more sophisticated investigations.

An excellent example of the progression of objectives in relation to an expanding conceptual base is provided by the work of Jungerius, Imeson, Duysings, Van Hooff, and Vis and their students and collaborators within the Keuper region of Luxembourg near Ermsdorf (cf. Study 42). Originally work focussed on six instrumented basins, but more recently attention has focussed on one completely forested basin. The initial objective was to provide general information on the water balance and on sediment and solute sources in the catchments. The second phase consisted of trying to relate channel and hillslope erosion processes to catchment outputs. The third phase is primarily concerned with integrating fluvial and hillslope processes with the dynamics of the ecosystem. Throughout these studies emphasis has been placed on the interaction of biotic and abiotic processes. The detailed objectives of some of the current investigations being undertaken by these workers also demonstrate the advantages of an interdisciplinary and integrated approach. They include:

a) Tracer investigations of the mechanisms of material transport on wooded slopes (Kwaad and Mucher);

b) The influence of Lumbricus on exposed areas of the forest floor and its significance for splash erosion (Van Hooff);

c) The role of present-day geomorphological processes in cuesta development (Jungerius and Van Zon);

d) Comparison of the sediment production from wooded and non-wooded watersheds (Imeson and Vis);

e) Establishment of process budgets (Duysings);

f) The geochemistry of weathering processes in the Keuper Shales (Verstraten and Dopheide);

g) The effects of overland flow on exposed forest floors (Imeson and Hendriks);

h) The rate of soil loss and colluviation in second-order drainage basins (Jungerius and Van Hooff);

i) The chemistry of atmospheric precipitation, throughfall and soil water under forests (Verstraten and Bouten).

Although model development clearly constitutes a longer-term objective of many current investigations, most are approaching that goal by means of improved understanding and parameterization of the processes involved. Few studies have been expressly designed to collect data for testing and developing existing models, although the Goodwin Research Catchment in northern Mississippi, USA provides a valuable example of such work. In justifying the establishment of this investigation, De Coursey (1982a) points to the recent development of models of soil erosion and deposition and solute generation such as CREAMS (Knisel, 1980) and SWAM (De Coursey, 1982b) and emphasises that the most significant problem associated with further development of these and similar models is probably the data base required to develop and test them. He contends that in many respects model development has progressed to the point that further improvement awaits data sets that can be used to assess improvement over previous versions or models.

2.4 CATCHMENT SCALE

The instrumented basins listed in Table 2.1 represent a wide range of scales. They range from the micro-watersheds of 10.6-44.3 m^2 in the badlands of southern Alberta investigated by Bryan and Hodges (1984) (Study 3), to large drainage basins in excess of 1000 km^2. In an attempt to represent the distribution of sizes associated with the seventy five studies, Figure 2.2 presents a frequency distribution of the areas involved. Where a study embraces a number of catchments, the average size of these has been incorporated in the distribution and where a sequence of sub-basins within a larger basin has been employed, the latter value has been used. With these limitations, Figure 2.2 suggests that the instrumented catchments currently being employed by geomorphologists are typically between 0.1 and 100 km^2 in extent.

Many factors govern the choice of catchment size, and the range of eight orders of magnitude evident in Figure 2.2 is in large measure a reflection of two opposing tenets: 'small is beautiful' and 'large is lovely'. In many circumstances it can be argued that a small basin provides the ideal unit of study, since it will afford control of major variables such as rock type and land use, it will permit intensive instrumentation, and it will minimize operational problems associated with reading or servicing a large number of instruments in a short space of time. Furthermore, in the case of the micro-catchments employed by Bryan and Hodges (Study 3) the use of a portable rainfall simulator provided a major constraint on the maximum feasible size. As a counter to these arguments, those workers employing larger basins could justify their choice of size in terms of the increased representativeness of a larger area and the fact that nature is typified by heterogeneity rather than by homogeneity. This is the disfunctional problem highlighted by Church (1984). Furthermore, it is known that catchment scale exerts an important influence on certain geomorphological processes such as sediment delivery

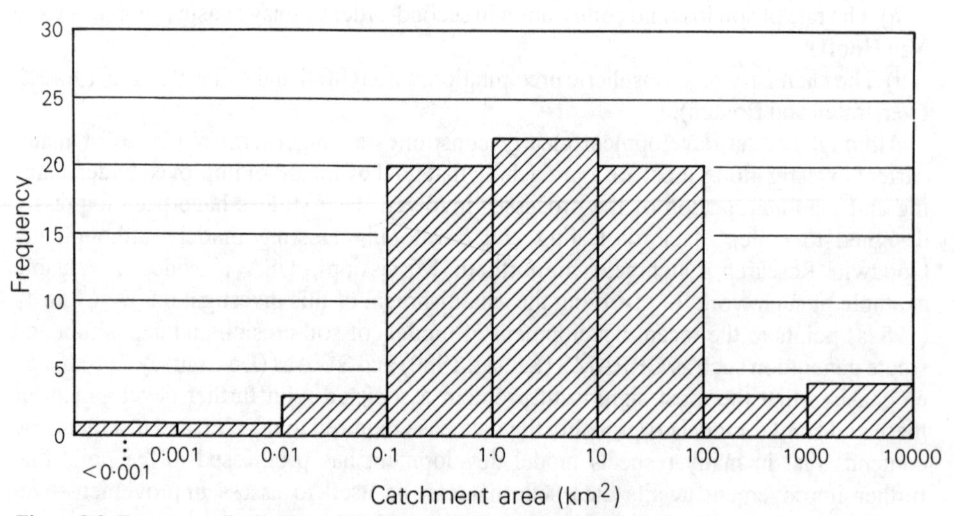

Figure 2.2. Frequency distribution of drainage basin area (from Table 2.1).

(cf. Walling, 1983), and it would be inappropriate to restrict studies of sediment budgets to small watersheds. Turning to the list of five major objectives embraced by this sample of catchment studies that was introduced earlier, it could be suggested that objective 1 has frequently been approached by using relatively large basins that provide results that are representative of a large area, whereas objectives 2-4 are in most cases better accomplished in a smaller basin. In the case of objective 5, catchment scale will clearly be governed by the character of the model. The large number of intermediate and large basins included in this sample of 75 studies is nevertheless somewhat surprising in view of the many advantages of small catchments for intensive process investigation.

Some of the problems of reconciling the advantages of large and small basins can clearly be overcome by the use of an hierarchical or nested approach. In this, provision would be made for study of small constituent basins within the larger unit. This is the strategy adopted in the Goodwin Creek investigation (Study 9), and in which both tandem and nested configurations were employed (cf. Fig. 2.8), in order to isolate major tributaries, to measure at least 60% of the intervening area of the 2.14 km^2 main basin, to provide subcatchments with relatively homogeneous soils, land use and geology, and to isolate significant sediment source areas and channels of different stability. In all, 12 gauging stations were established. Further examples of this approach are provided by the work of Ambroise and his colleagues (Study 47) in which the Ringelbach basin (36 ha) constitutes part of the 12.3 km^2 Petite Fecht basin and this in turn is a sub-basin of the 450 km^2 Fecht catchment; by the Nahel Yael watershed investigated by Schick and his collaborators (Study 57); and by the Homerka catchment instrumented by Froehlich (Study 32). In the latter case, the 19.55 km^2 Homerka catchment was further subdivided into the 3.95 km^2 Bacza sub-basin and an 'experimental slope' with detailed instrumentation of runoff and sediment contributions from unmetalled roads and fields (cf. Fig. 2.7). At a different scale, the work of Walling and Webb in investigating sediment and solute delivery and transport by the River Exe and its major tributaries (Study 24) involved instrumentation of major tributaries, some intermediate mainstream locations and a number of sub-basins (cf. Fig. 2.9).

2.4.1 *Lake, reservoir and glacier catchments*

In three cases, namely, Studies 19, 61 and 62, the choice of study catchments has clearly been governed by the presence of a reservoir or lake at the catchment outlet. Surveys of reservoir sedimentation have been used in these investigations to provide estimates of sediment yield. Such estimates have the advantage of including both the suspended and bed load components of total load and of providing a mean value representative of a longer period of time than can commonly be obtained using direct measurements of sediment transport. However, this approach can only provide a temporally-lumped view of sediment yield since it is impossible to monitor the deposition associated with individual events. The detailed studies of sediment cores being undertaken by Foster and his colleagues in the Merevale Lake Study (Study 19) do, nevertheless, offer very considerable promise for deciphering long-term trends in sediment yield and in the relative importance of major sediment sources, in response to changes in catchment condition. The potential of lake-catchment studies for geomorphological investigations will be reviewed further later in this report.

Studies 2, 27, 28 and 29 are also distinctive in that they are concerned with drainage basins including a glacier. As such they provide a valuable means of assessing the influence of the glacier on denudation processes both beneath the ice and in the proglacial area.

2.4.2 *Methods*

There can be little doubt that the technological advances of recent years have resulted in major developments in the range and sophistication of the instrumentation available for geomorphological investigations. Perhaps the greatest progress relates to the possibility of replacing a programme of intermittent manual recording by automatic continuous recording equipment, although the value of such recorders depends heavily upon the parallel development of appropriate sensors and transducers. Instrumentation of catchment studies has therefore moved a long way since the days of the vigil network (cf. Leopold, 1962; Leopold and Emmett, 1965), when the emphasis was on simple, cheap and durable monitoring devices, although it is not easy to state whether it is the availability of more sophisticated instrumentation that has permitted the investigation more demanding problems, or whether the increasing complexity of the questions posed by geomorphologists has itself generated these developments. It is, however, important to recognise that there is not necessarily a direct relationship between the degree of sophistication of the instrumentation involved in a catchment study and the value and significance of the results obtained. Furthermore, the availability of sophisticated instruments does not remove from the geomorphologist the need for careful formulation of the problem to be investigated and of the approach to be followed.

In reviewing in more detail the methods employed in the sample of 75 catchment studies included in Table 2.1, an attempt has been made to highlight recent developments in instrumentation and to point to promising techniques which might usefully be included in future investigations, rather than to provide an exhaustive survey of all methods employed. In this context it is suggested that basic instrumentation such as flow measuring weirs and water level recorders can be viewed as standard equipment and do not merit inclusion. Significant developments include the following:

Recording and logging equipment

Magnetic tape loggers and loggers with solid state memories have provided a major breakthrough in the potential for the continuous and synchronous recording of a wide variety of parameters in catchment studies. Studies 4, 18, 47 and 67 report the routine use of magnetic tape loggers for data collection and solar panels have been used in several of these studies in order to permit their operation at remote sites. In the Goodwin Creek study (Study 9), a more sophisticated system involving direct transmission of data from field sites to a central data collection facility by VHF radio telemetry is employed. Each field monitoring station, which collects data on climatological parameters, rainfall, streamflow, sediment transport and water quality is controlled by a microprocessor. This interrogates up to 16 individual sensors at a present frequency of between 1 and 30 minutes and stores the data in a 3K random-access memory. These data are transferred to the Central Control Computer at the laboratory headquarters by radio telemetry every 30 minutes. This central control system is able to interrogate up to 30 remote data stations during each 30 minute period. The incoming data are quality controlled and subsequently stored on disc file in 100 character records. As such, this system represents near real-time data collection, although the costs and technical expertise involved clearly place the approach beyond the scope of most investigations.

Increasing attention to the measurement of runoff processes has necessarily focussed attention on the need to monitor spatial and temporal variations in soil moisture and more particularly the dynamics of saturated areas. Traditional hydrological techniques, including piezometers, tensiometers and neutron probes have been widely applied, as evidenced by Studies 13, 16, 18, 20, 21, 26, 47, 65 and 67. The automatic scanning tensiometer system described by Anderson and Burt (1977), which has been used effectively in Studies 15, 20 and 21, provide a useful means of recording the response of a network of tensiometers, using a single pressure transducer and recorder. Various methods for monitoring the flow of water through soil pipes have also been developed by Jones in the Maesnant catchment in Wales, UK (Study 18).

Measurement of soil water chemistry

Several studies of soil water quality have also been undertaken as a means of studying patterns of solute generation and associated chemical denudation on hillslopes. The use of throughflow pits to study the chemistry of seepage from individual soil horizons at different locations is reported by Williams and his colleagues working in the Narrator catchment in Devon, UK (Study 25), and by Studies 13 and 15. Vacuum soil-water extractors (e.g. Parizek and Lane, 1970) have also been successfully used to sample soil water in a number of studies including 1, 5, 13, 15, 19 and 67, and this technique offers considerable promise as a relatively easy means of frequently sampling soil moisture under saturated and unsaturated conditions with a minimum of soil disturbance.

Dye tracing of subsurface flow

The use of dyes to trace subsurface flow has been proposed by a number of workers in recent years (e.g. Smart and Laidlaw, 1977) and this technique has also been applied in a number of geomorphological studies. Trudgill and his co-workers in the Whitwell Wood Investigation (Study 15) report the successful use of the fluorescent dye Lissamine FF to trace the movement of precipitation inputs though the soil to the stream. They were, for example, able to demonstrate that stemflow, which may represent a point source input of large volumes of water, bypassed the soil matrix by flowing along root channels and macropores and into the fissured bedrock without diffusing into the soil. In the study of the Hachioji drainage basin, Japan, undertaken by Tanaka and his colleagues (Study 65) Sulpho Rhodamine B and Lissamine FF have been used to measure the velocities of pipeflow. Dyes have also been successfully used to trace glacier-groundwater interactions by Smart working in the Rocky Mountains of Alberta, Canada (Study 2).

Use of micro-weight loss rock tablets to document spatial variations in chemical denudation rates at the soil/rock interface

The successful use of rock tablets to study spatial variations in chemical denudation rates across the slopes of a drainage basin has been reported by workers in the Bicknoller (Study 21) and Whitwell Wood (Study 15) catchments in the UK. Although this micro-weight loss approach does involve problems in terms of the representativeness of the results in absolute terms, it nevertheless provides valuable relative data and an effective means of testing theoretical models of slope development. Thus results from the Whitwell Wood investigation have shown that solution rates at the soil-bedrock interface increase upslope and that slope development by this process will ultimately lead to slope decline. In the Bicknoller study, the results obtained were related to detailed information on soil moisture conditions on the slope obtained from tensiometer measurements and in this case it was shown that maximum denudation rates occurred within a hillslope hollow which was the location of a saturated wedge. Scope clearly exists for comparing values of denudation rates obtained using this approach with values estimated from measurements of solute yield at the catchment outlet.

Automatic water samplers

Measurements of the output of sediment and solutes from a drainage basin form a major part of many catchment studies and the development of reliable automatic water sampling equipment represents a major advance. This equipment enables samples to be collected at frequent intervals during storm events, and therefore provides the basis for accurate assessments of material outputs. Automatic samplers are now employed in the majority of studies attempting to evaluate detailed solute budgets (e.g. Studies 9, 14, 15, 19, 23, 25, 26, 27, 69, 70 and in many sediment yield studies (e.g. Studies 4, 6, 8, 17, 19, 23, 28, 56, 57, 68, 69, 70).

Continuous recording of suspended sediment concentrations

The rapid fluctuations in suspended sediment transport encountered in many streams have also encouraged attempts to develop methods for continuous recording of sediment concentrations. Photoelectric turbidity meters have been successfully employed for this purpose in Studies 17, 19, 23, 24 and 42, and although careful attention must be given to calibration problems (cf. Fish, 1983) there can be little doubt of the value of such equipment. Photoelectric measurements are, however, effectively restricted to sediment concentrations less than 10000 mgl^{-1} and alternative approaches must be employed where higher levels are encountered. Tazioli and his colleagues working in two catchments in southern Italy (Study 54) have obtained good results from radioactive sediment concentration gauges utilising ^{241}Am and ^{137}Cs sources. In the former case, the measuring gauge contains both the ^{241}Am source and the gamma scintillation detector and is mounted in the channel, whereas in the latter instance the source and detector are permanently installed within the walls of the gauging structure and on opposite sides of the channel. Encouraging results have also been obtained from preliminary trials of vibrating U-tube density cells for continuously monitoring suspended sediment concentrations in the Goodwin Creek investigation (Study 9). With this apparatus water is pumped from the stream through the vibrating U-tube which produces an electrical output proportional to sediment density.

Continuous monitoring of specific conductance

Continuous measurements of specific electrical conductance have been profitably employed in a number of geomorphological studies of solute transport. Walling and Webb report the extensive deployment of specific conductance recording equipment in the Exe basin in Devon, UK (Studies 23 and 24), and Andrews has used similar equipment in his study of sediment and solute transport in the Piceance basin, Colorado, USA (Study 8). For most rivers specific conductance values can be readily converted to estimates of total dissolved solids concentration, and in some studies well defined relationships between specific conductance and cation concentrations have also been established. Collins describes the effective use of a specific conductance record to estimate cation concentrations in the meltwater stream of the Gornera glacier in Switzerland (Study 27).

Bed load measurement
The accurate measurement of bed load transport remains a major problem in most investigations of material transport from drainage basins. The Helley-Smith sampler developed by the US Geological Survey (cf. Emmett, 1979) has been used with encouraging results in Studies 7, 17, and 74 and this sampler may merit trials in other geomorphological studies. Construction of a large detention basin to trap the total bed load transport from a catchment has proved possible in the Nahel Yael study of Schick and his co-workers (Study 57) and in the investigations undertaken by Griffiths in the Dry Acheron basin in South Island, New Zealand. However, this strategy would prove impossible in many studies due to the lack of a suitable site for the excavation. The use of a large concrete-lined pit trap installed within the channel and emptied after each flood event using a mechanical excavator has been reported from the work of Newson and his colleagues in the Plynlimon catchments in Wales, UK (Study 17), and may provide a satisfactory alternative to the detention basin in many situations. Vortex bed load traps have also been used with very good results in the Torlesse catchment in New Zealand by Hayward (Study 75) and by Tacconi and his co-workers in the Virginio basin in Tuscany, Italy (Study 53). Although demanding a more complex installation than a simple pit trap, the vortex trap possesses the advantage that a near continuous record of the magnitude and calibre of bed load transport can be obtained if the outlet from the vortex tube is coupled with a conveyor belt system to handle the bed load discharged.

Analysis and interpretation of suspended sediment properties
Whereas geomorphological studies of suspended sediment transport have traditionally focussed on assessing the magnitude of the load transported, a number of recent studies have indicated that it may also be profitable to direct attention to the properties of the sediment transported. Walling has used information on the mineralogy, magnetic properties and organic and inorganic chemistry of suspended sediment to 'fingerprint' sediment sources in the Jackmoor Brook and Dart catchments in Devon, UK (Study 23) and also to evaluate the selectivity of the erosion and conveyance processes. Such work will, however, demand a new approach to sediment sampling in many areas because of the problems of obtaining suitable amounts of sediment (e.g. 20 g) from a river with relatively low concentrations of suspended sediment (e.g. 500 mgl^{-1}). Walling successfully employed a continuous flow centrifuge to separate sediment from bulk samples of river water obtained from rivers in Devon, UK (Study 23).

Studying soil loss and sediment redistribution using ^{137}Cs
Although the potential for using measurements of the ^{137}Cs content of soil profiles to elucidate spatial patterns of soil loss and redistribution within a drainage basin has been recognized for several years (e.g. Ritchie et al., 1974), geomorphologists have as yet made relatively little use of this approach to studying erosion and deposition. It is, however, interesting to note that Loughran and Campbell have employed this technique to study sediment movement within the Maluna catchment in New South Wales, Australia (Study 68) with considerable success. They were able to investigate the relative

importance of cultivated and uncultivated areas as sediment sources by comparing the ^{137}Cs content of representative soil profiles and to identify areas of sediment storage such as small alluvial fans by studying both the ^{137}Cs content of soil cores and the vertical distribution of ^{137}Cs within the core. Since ^{137}Cs inputs to the landscape commenced with the onset of atomic weapon testing in 1955, this environmental isotope could provide the geomorphologist with a valuable means of studying landscape development over periods of tens of years and therefore of extending shorter term process measurements.

2.5 INSTRUMENT NETWORKS

In reviewing the methods currently being used in catchment investigations, it is important also to consider the overall schemes of instrumentation and the associated networks that are being employed in individual studies. These will depend upon the scale and objectives of the study and may be illustrated by considering in more detail a representative selection of the listed investigations.

2.5.1 *The Hachioji basin (Study 65)*

This small (0.022 km^2) forested basin near Tokyo, Japan is being studied by a research group from the Institute of Geoscience of the University of Tsukuba. The main objective of the investigation is to study the runoff generation processes operating within the basin and to relate these to processes of soil erosion. A special feature of the drainage basin is the extensive valley floor (Fig. 2.3A) and the instrumentation has been concentrated into this area, since it represents the most active zone in terms of the incidence of surface runoff and associated erosion.

To analyse the dynamic response of the basin during a storm event, a dense network of tensiometers and piezometers has been employed (Fig. 2.3B). Piezometer nests of one to three piezometers have been installed at 12 different sites and 10 tensiometer nests comprising a total of 54 tensiometers have also been installed. Output from the basin is measured by a 90° sharp-crested V-notch weir where suspended sediment samples are also collected. Analysis of data from the piezometer and tensiometer networks and field mapping of saturated areas during storm events permits detailed evaluation of areas contributing to surface runoff and of the runoff processes involved (Fig. 2.3C). These results suggest that the majority of the overland flow comprises return flow appearing at the surface through decayed stumps and soil pipe outlets concentrated in three major zones of the valley floor. Further studies have been undertaken to trace the movement of water through the soil pipes. Variation in suspended sediment concentration at the measurement weir have been related to the relative importance of surface and subsurface flow contributions and a Subsurface Water Exfiltration Erosion Model (SWEEM) has been proposed. This study therefore provides a useful example of the instrumentation associated with a study where an examination of runoff processes is being used to improve understanding of the erosion processes operating in the basin. Other examples of similar work include the study of Anderson and Burt in the Bicknoller catchment, UK (Study 21) which is particularly notable in terms of instrumentation in that an automatic scanning tensiometer system was used.

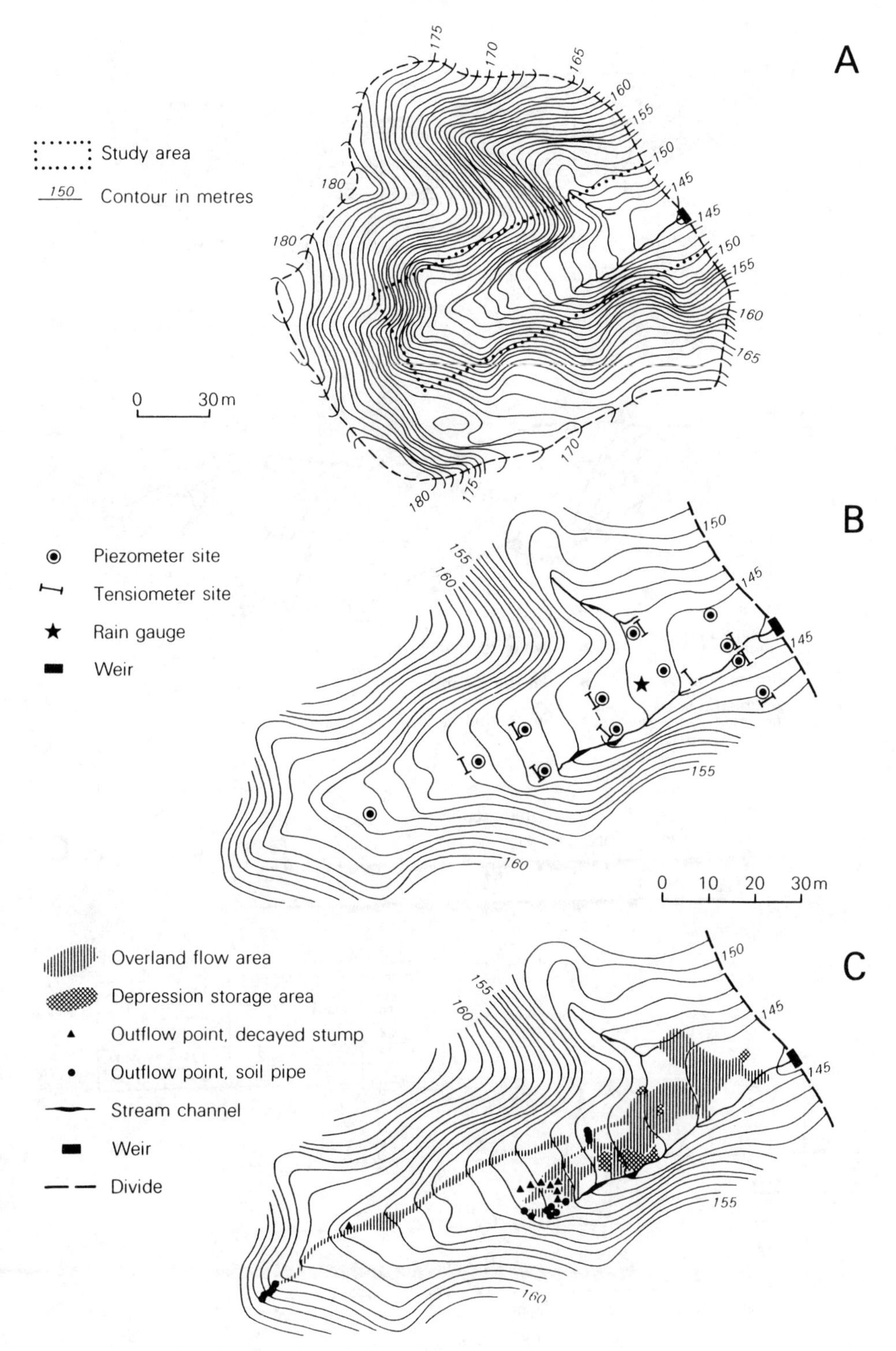

Figure 2.3. The Hachioji basin study (based on Tanaka, 1982).

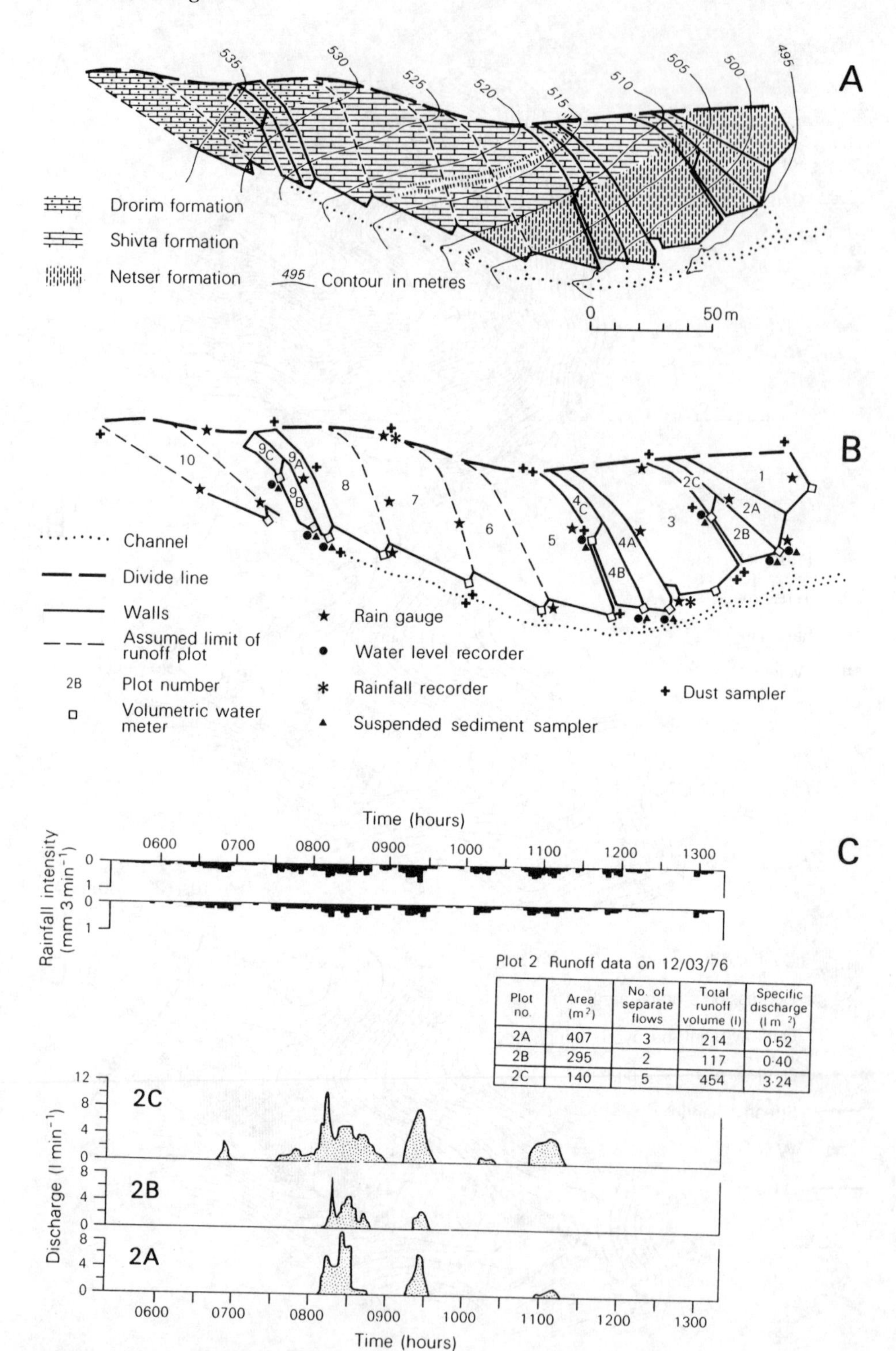

Plot no.	Area (m²)	No. of separate flows	Total runoff volume (l)	Specific discharge (l m⁻²)
2A	407	3	214	0·52
2B	295	2	117	0·40
2C	140	5	454	3·24

Figure 2.4. The Sede Boker study (Yair, 1981).

2.5.2 *The Sede Boker investigation (Study 56)*

This investigation undertaken by Yair and his colleagues in the northern Negev, Israel, provides another example of a study of runoff and sediment generation processes. In this case, however, the environment is arid (mean annual rainfall 93 mm). Again emphasis has been placed on illuminating the spatial variability of runoff and sediment production, and the study has focussed on the north facing slope of a small drainage basin underlain by three lithological formations (Fig. 2.4A). A series of ten contiguous plots draining parallel strips of hillslope have been established (Fig. 2.4B) and three of these plots (Nos. 2, 4 and 9) have been subdivided into three subplots representing a unit extending from the divide to the slope base, and two adjoining short units draining the upper and lower sections of the slope. This configuration permits an analysis of spatial variations in runoff and erosion according to different lithological units and combinations of lithology, and to different positions on the slope (upper and lower). Nine of the total of 16 plots are equipped with runoff measuring devices and automatic suspended sediment samplers, whilst the remaining plots are equipped with volumetric water meters. A dense network of rain gauges has also been established to take account of the marked spatial variations in rainfall totals that may occur even over a small area such as this.

Figure 2.4C provides an example of the data collected from this instrumentation network. It illustrates the rainfall recorded at two recording gauges and the runoff records from plots 2A, 2B and 2C. The spatial variability of runoff response is clearly demonstrated by these records. The upper part of the slope (Plot 2C) generates considerably more runoff than the lower slope (Plot 2B) and the data from the complete slope (Plot 2A) suggest that most of the runoff originating upslope is absorbed, before reaching the channel, by the colluvial material covering the lower part of the slope. Measurements of sediment production obtained from these plots again demonstrated marked spatial variation, with plots 7, 8 and 9 contributing the majority of the total sediment yield from the instrumented slope over a one year period. This was explained in terms of the higher runoff production from the massive limestone of the Shivta formation and the high biological activity found on this formation, which is associated with sediment availability. Burrowing and digging activity by desert animals such as isopods (*Hemilepistus reaumuri*) provides a major sediment source in this environment.

2.5.3 *Lone Tree Creek basin (Study 7)*

Turning to an example of a small catchment study where an attempt has been made to establish a complete sediment budget using relatively simple techniques, it is instructive to consider the work of Lehre in the 1.74 km^2 Lone Tree Creek basin in the Coast Ranges of north-central California. This study has attempted to define the quantitative relationship between sediment mobilization, sediment output and associated changes in storage within the basin. Some details of the conceptual model produced to represent the general linkages between sediment storage sites and erosional processes and transfers between storage elements are shown in Figure 2.5. The methods employed to document sediment mobilization (the total amount of sediment moved any distance by a process), sediment production (the amount of mobilized sediment reaching a channel), and sediment yield (the amount of sediment actually discharged from the drainage basin) are listed in Table 2.2.

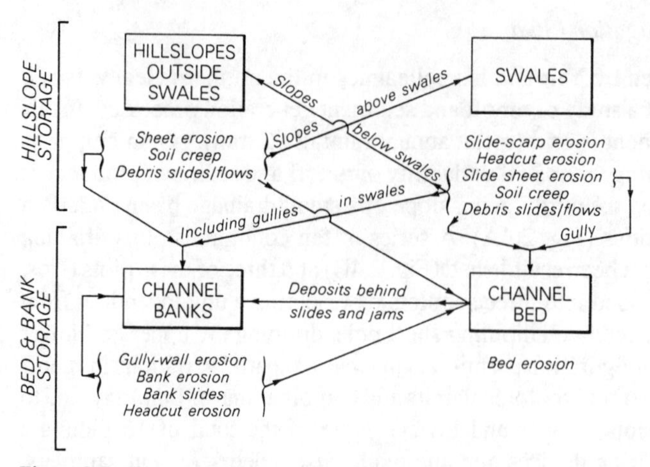

Figure 2.5. The conceptual model of linkages between sediment storage sites and erosional processes developed by Lehre (1982) in his study of the Lone Tree Creek catchment, California. Boxes indicate storage elements; listed below each box are erosional processes mainly responsible for mobilizing sediment in that element. Arrows show transfers between elements. Labels on arrows qualify or restrict location of transfers (based on Lehre, 1982).

Table 2.2. Methods employed for documenting the sediment budget of the Lone Tree Creek basin. Based on information in Lehre (1982).

Process	Methods
Landslides	Field mapping, sequential comparison of aerial photos
Slide-scarp erosion	Survey stakes
Slide-scarp sheet erosion	Nail and washer erosion pins
Headcut retreat	Survey stakes
Soil creep	Creep test-pillars
Hillslope sheet erosion	Erosion pins
Gully scarp erosion	Survey stakes
Bank erosion	Channel traverses and mapping of bank slides
Bed erosion	Surveyed cross sections
Suspended sediment yield	Manual sampling
Bed load yield	Manual sampling (Helley-Smith Sampler)
Dissolved load yield	Manual sampling

2.5.4 *The Dinosaur Park catchment (Study 4)*

A more detailed attempt to study the relationship between on-site erosion and downstream sediment yield, and the associated conveyance or delivery processes, is represented by the work of Campbell and Bryan in a small 0.36 km^2 catchment developed in the badlands of the Dinosaur Provincial Park in Alberta, Canada. These workers are investigating the relationships between microscale processes, which they have already studied using sprinkler experiments (Study 3), and the response of a mesoscale badland basin. Seven subbasins have been defined within the catchment in order to represent the

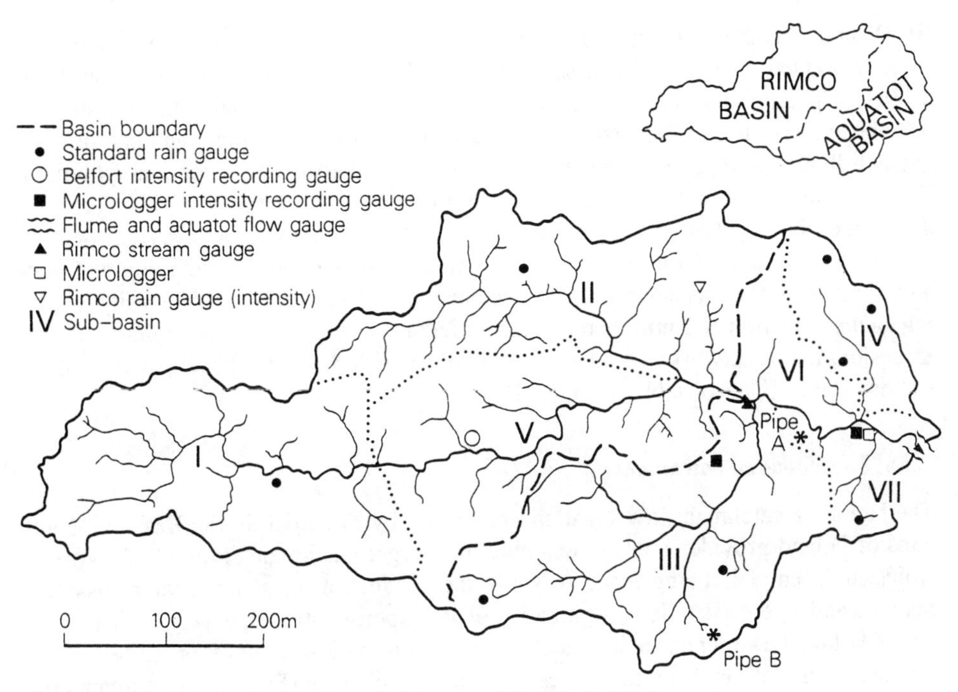

Figure 2.6. The Dinosaur Park basin study (Bryan and Campbell, 1982).

Table 2.3. Characteristics of the sub-basins included in the Dinosaur Park catchment.

	Area (m^2)	Sub-basin	Area (m^2)	Surface characteristics
		I	72340	Gently sloping, largely grass-covered, few high-yielding units
Rimco Basin	202250	II	68470	Gently sloping pediments and sandstone
		V	61450	Gently sloping, extensive grass cover, piped-sandstone near divide
		III	79230	Gently sloping pediments, steep piped shales
Aquatot Basin	134550	IV	15620	Steep shales and sandstone, much instability
		VI	23760	Extensively piped irregular shales
		VII	10600	Steep bare shales with sandstones near divide
	Undefined areas		5340	
Total			336810	

major geomorphic and lithologic units in the area (Fig. 2.6, Table 2.3). Thus, for example, subbasin I contains flattish sandstone pediments and an extensive flat grassed surface, subbasin IV is a narrow, deeply-incised valley with steep slopes with abundant mass movement features, and subbasin VI is a poorly-defined area of complex morphology honeycombed by pipes and tunnel erosion. In terms of routing processes, subbasins II,

III, IV and V connect directly to the main channel, subbasin I enters the system in a sinuous braided channel through subbasin V, and a large proportion of the flow from subbasins VI and VII drains through a diffuse pipe network. Two major gauging stations have been constructed. The first is at the outlet of the total catchment and comprises a Parshall flume equipped with a sonic flow gauge (Aquatot). The second, employing a Rimco stream gauge is above the confluence of subbasins III and IV. Both incorporate data loggers powered by solar panels and Isco automatic water samplers. A Helley-Smith sampler has been used for bed load measurements. A network of recording and non-recording precipitation gauges has also been installed. The micrologger intensity recording gauges function at 1 mm increments and record on a time base of 1 minute. Future extensions to this network which have been planned include the establishment of gauging stations at the outlets of subbasins I and II.

2.5.5 *The Homerka catchment (Study 32)*

The Homerka catchment investigation undertaken by Froehlich in the Beskidy Mountains of Poland provides a useful example of a larger catchment study (19.55 km^2) in which an attempt is being made to study the production of sediment and dissolved material and where a detailed investigation of an 'experimental slope' (Fig. 2.7) has been nested within the larger basin, in order to provide information on runoff processes and the movement of material from the slopes to the stream channel (Fig. 2.8). Instrumentation

Figure 2.7. Experimental slope in Homerka basin (Froehlich, 1983).

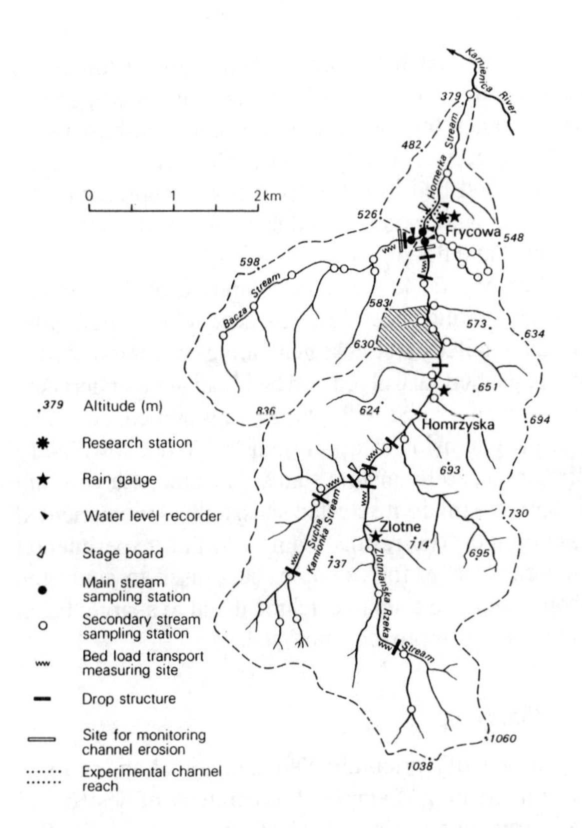

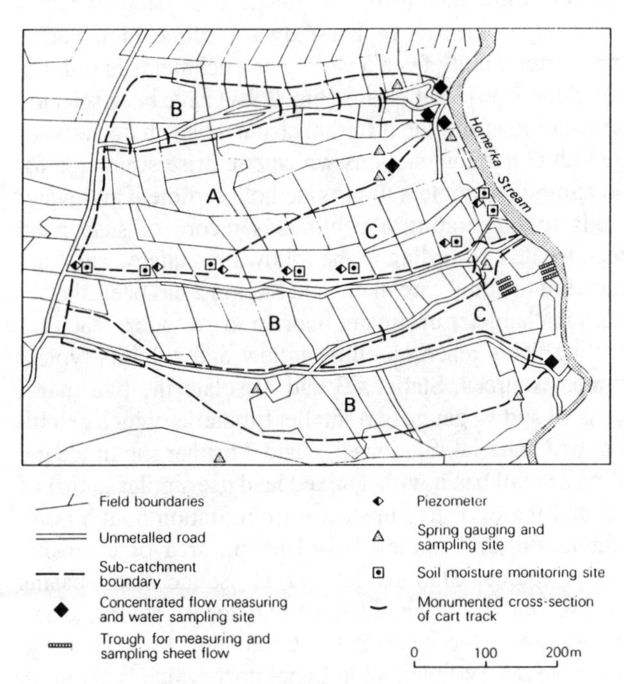

Figure 2.8. The Homerka basin study (Froehlich, 1983).

within the larger basin and the deforested sub-catchment of the Bacza Stream comprises a hydrometric network of precipitation gauges, water level recorders and staff gauges; sediment and solute sampling sites; reaches for measuring bed load transport using painted pebbles; an experimental channel reach for conducting detailed investigations of the mechanisms of suspended sediment and bed load transport (e.g. longitudinal and lateral dispersion of suspended sediment); and monumented sections for documenting erosion and deposition within the channel and flood plain.

The 'experimental slope' has an area of 0.27 km^2 and is composed of three major subunits (Fig. 2.8). Firstly, the slopes generating sheet flow (C), secondly, a small gully watershed (B), and thirdly, a series of unmetalled cart tracks generating concentrated flow (A). Measurements of runoff and of suspended, dissolved and bed load were undertaken for each of these subunits. Sheetflow was measured with plastic traps and concentrated flow in the gully and cart-tracks was gauged using sharp-crested 90° V-notch weirs. In addition measurements of spring discharge, soil moisture status, piezometer levels and the chemical composition of soil water were made at selected sites, and 30 monumented sections on the cart tracks were monitored. The results from both the 'experimental slope' and the larger Homerka catchment have, for example, been used to study the problems of extrapolating results from a slope to a small catchment and to a larger basin and have demonstrated the value of the nested approach employed.

2.5.6 *The Goodwin Creek catchment (Study 9)*

This study undertaken by the US Department of Agriculture Sedimentation Laboratory at Oxford, Mississippi, USA, affords an interesting example of a hierarchy of nested and tandem subbasins which have been instrumented using a sophisticated system of radio-telemetry. As outlined previously, a major objective of this study is to collect high quality data on runoff, sediment yield and water quality from a variety of catchments in order to test and develop mathematical models. The 13 subbasins (Fig. 2.9A) have been selected to represent the various land use conditions and the different channel bed characteristics within the larger basin. The individual gauging stations which comprise structures for measuring discharge, automatic pumping samplers to provide both sediment and water quality samples and density cells for generating a continuous record of suspended sediment concentration have been located according to the following strategy. Station 1 provides measurements at the outlet of the total catchment and Station 2 has been sited at a point where the channel changes in character upstream, in order to provide a reach for studying flow and sediment routing. This reach has little inflow and exhibits typical channel configuration and sediment sources. Stations 3 and 4 isolate the two major tributaries of the basin and Stations 13 and 14 gauge two smaller tributaries which exhibit a much larger size and amount of bed material. Stations 5, 6 and 7 further subdivide the catchment, with Station 6 isolating a small basin with a mixed land use similar to that of the whole catchment. Stations 8 and 9 are located upstream from Station 5, at a point where the basin changes configuration, and Station 9 isolates an area of extensive gullying that is a major sediment source. Stations 10 and 11 isolate small basins characteristic of wooded and pasture land use respectively and Station 12 is located at a major change in channel configuration at the head of the catchment. Data from the gauging sites are transmitted to the USDA Sedimentation Laboratory which is about 38 km away by a radio-telemetry system, described in a previous section of this report, that

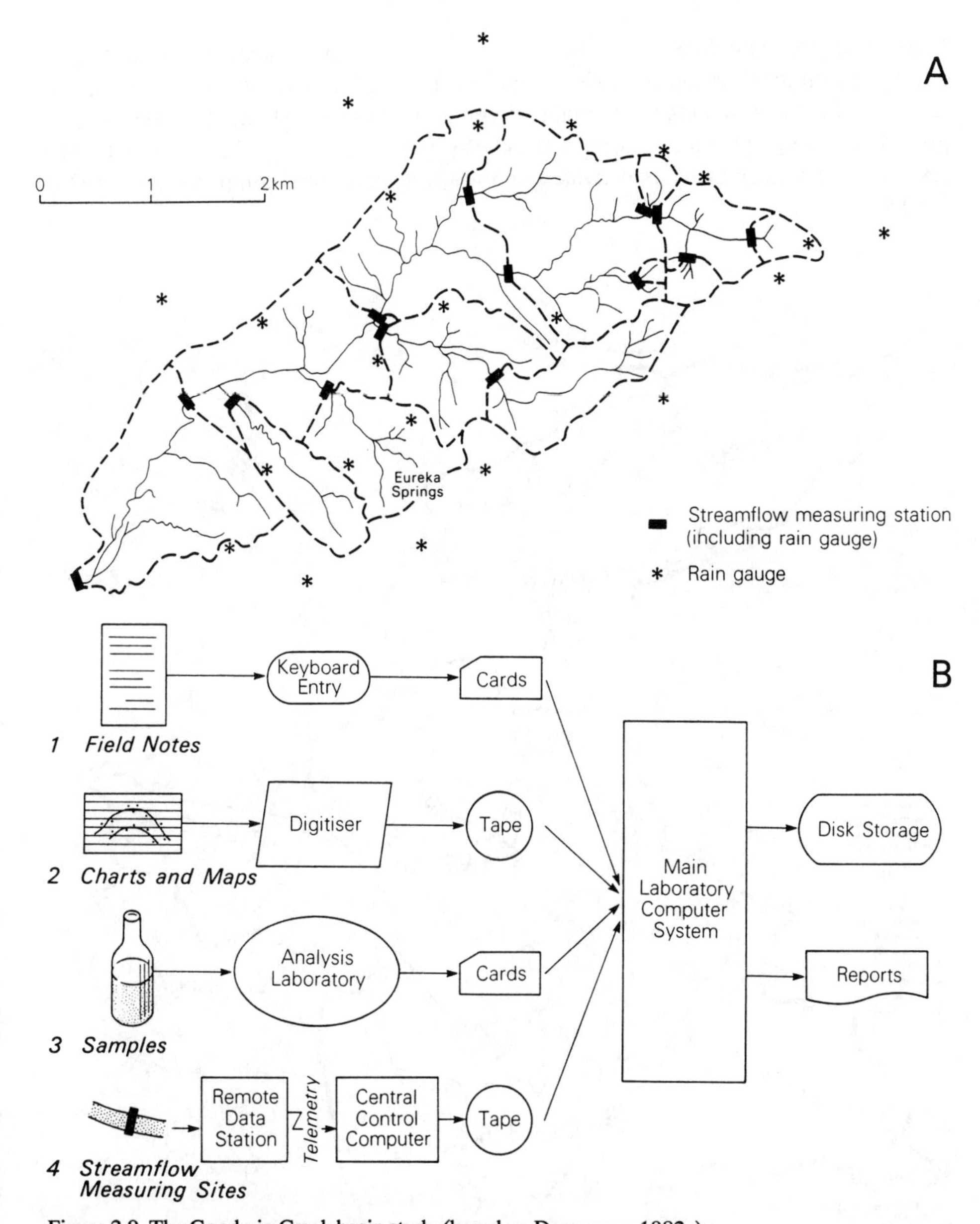

Figure 2.9. The Goodwin Creek basin study (based on Decoursey, 1982a).

affords near real-time data availability, time synchronization and the ability to select a high or low frequency of observation.

In addition several small source area watersheds have been established within the basin to provide detailed information on such hydrologic parameters as roughness and canopy cover. These are operated for short periods of time (1-2 years) and will be moved from site to site to produce a wide range of values. An intensive network of rain gauges and a climatological station also exist. In addition to the use of radio-telemetry to transmit data from field instruments to the main laboratory computer system, attention has been given to the development of a comprehensive system for entering all data, including the results of laboratory analysis of water samples, into the same computer system (Fig. 2.9B).

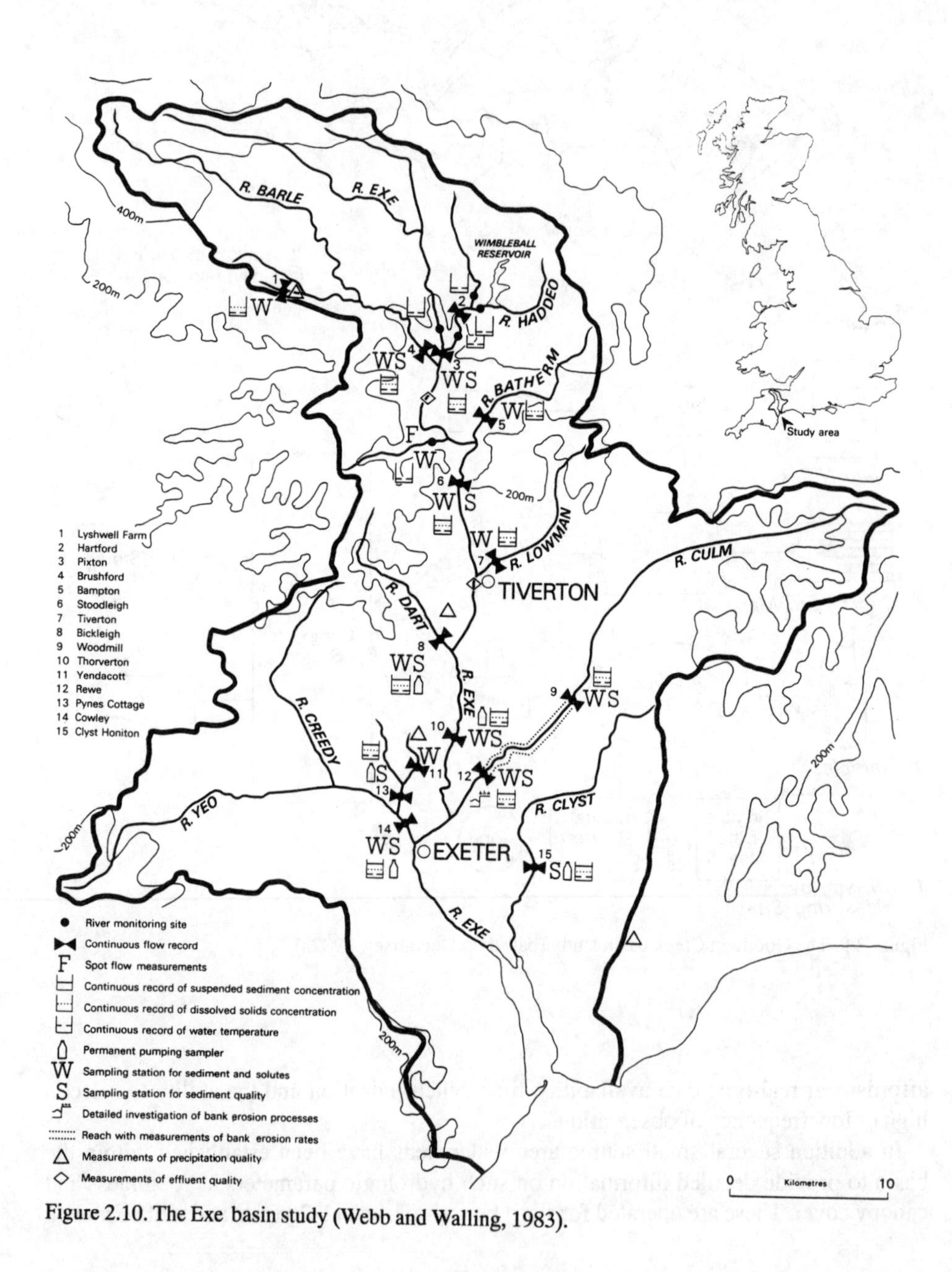

Figure 2.10. The Exe basin study (Webb and Walling, 1983).

2.5.7 *The Exe Basin study (Study 24)*

This study initiated by Walling and Webb in Devon, UK provides a final example of catchment instrumentation that relates to an investigation of spatial variations in suspended sediment and solute yield over a wide area (c. 1500 km^2) and of sediment conveyance through a river network. The Exe Basin exhibits a wide range of terrain types which vary from unenclosed moorland developed on resistant Devonian slates and sandstones in the northern part of the catchment, through the dissected landscape of the Middle Exe Basin underlain by Carboniferous sandstones and shales, to intensively farmed lowlands in the south characterised by Permian breccias, sandstones and marls. This diversity of terrain is paralleled by a wide range of hydrometeoroiogical conditions, with values of mean annual precipitation and runoff varying from 1800 mm and 1500 mm respectively in the northern headwaters to less than 800 mm and 300 mm in the southern part of the catchment.

The 15 main measuring stations shown on Figure 2.10 have been located to fulfil two major objectives: firstly, to provide representative coverage of tributaries draining the various terrain types in the basin (i.e. Stations 1, 2, 3, 4, 5, 7, 8, 9, 11, 13, 14 and 15) and secondly, to generate information on the transmission of suspended sediment through the river network (i.e. Stations 1, 4, 6, 10, 12, 11, 13 and 14). In the first context, the catchments associated with Stations 1, 8 and 13 have been selected to provide locations for more detailed work on small/intermediate sized basins representative of the three major terrain types. In the second context, the river reach between Stations 9 and 12 provides the location for a more detailed investigation of sediment routing through a meandering reach with a well-developed floodplain.

The network of instrumentation (Fig. 2.10) makes extensive use of equipment for continuous monitoring of suspended sediment concentrations (photoelectric turbidity meters) and total dissolved solids concentrations (specific conductance monitors), and of purpose-built automatic pump sampling equipment. It also includes a large number of sites where manual sampling of suspended sediment and solute concentrations is undertaken and where bulk samples of suspended sediment are collected. The reach between Stations 9 and 12 is instrumented with erosion pins and surveyed sections to monitor channel erosion and deposition and with sediment traps to monitor overbank deposition. The storage of fine-grained sediment within the channel network is also monitored on a regular basis at a number of representative sites.

2.6 RESULTS

Any attempt to summarise the results of seventy-five catchment studies with diverse objectives, many of which represent the work of a team of researchers working over periods of a decade or more, and for which recent results may as yet be unpublished, faces a daunting task. The reader is referred to the volume containing the papers presented at the UK meeting of the Commission in 1981 on 'Catchment experiments in fluvial geomorphology' (Burt and Walling, 1984) for a representative sample of detailed results from ongoing studies. In this report, attention will rather focus on the question of 'what do we now know that we did not know 10 years ago?', although even the answer to this must inevitably be subjective and generalized. It is suggested that an answer can be structured to include a number of themes which will be considered in turn.

Firstly, there can be little doubt that we now possess an improved understanding and conceptualization of many of the contemporary processes operating within the drainage basin. A random sample of citations from the studies listed in Table 2.1 could refer to the pioneering studies of Anderson and Burt in the Bicknoller catchment (Study 21) and of Anderson and Kneale in the Winsford catchment (Study 20) which have improved our understanding of runoff processes and demonstrated the importance of topography in controlling these processes and, more particularly, the dynamics of saturated areas; to the work of Jones in the Maesnant catchment (Study 18) which has provided valuable information on the role of soil piping in runoff production in upland areas; and to the work of Yair in the Sede Boker study (Study 56) which has provided important evidence of partial area contributions to runoff from arid areas as distinct from humid regions. On the same basis, the work of Gallie and Slaymaker (Study 1), the investigation of Afifi and Bricker in the Mill Run basin in Virginia, USA (Study 10) and of Johnson and Likens and their co-workers in the Hubbard Brook catchments (Study 11), and the studies of Walling and Webb in the Exe basin (Studies 23 and 24) and Williams and Ternan and their colleagues in the Narrator catchment (Study 25), have provided the basis for a considerable improvement in our understanding of the processes and pathways involved in solute generation within a drainage basin, and of the factors controlling solute yields at a catchment outlet. Similar comments could be directed at other processes including the dynamics of erosion and sediment conveyance and it is to be hoped that this improved conceptual base will provide the foundation for the development of the more rigorous scientific experiments advocated by Church (1984).

Secondly, and closely connected with the above comments, the expanding spatial coverage of catchment studies, when viewed at the global scale, must be welcomed, since it is important that information concerning the operation of fluvial processes should be available from as wide a range of climatic and physiographic environments as possible. Ten years ago a similar review to this would have been almost entirely restricted in its spatial coverage to Europe, North America, and Israel, but now it is possible to point to the work of Nortcliff and Thornes (Study 13) and Netto (Study 12) as representing what are probably the first studies in South America, as well as to investigations in Nigeria, Zaire, Zimbabwe, Lesotho, South Africa, China, Japan, Java, Australia and New Zealand. To cite just three examples, information is therefore now available on the precise nature of runoff generation processes in tropical rainforest areas from the studies of Bonnell and his co-workers in northern Queensland, Australia (Study 67), on sediment and solute yields from small drainage basins representing a variety of environments in North and South Island, New Zealand from the work of O'Loughlin, Pearce and Rowe (Study 72) and on the exceptionally high erosion rates to be encountered in the loess region of the Middle Yellow River basin in China from the work of Mou and Meng and their colleagues from the Yellow River Conservancy Commission (Study 63).

There are still many areas of the world where data on the nature and rates of operation of fluvial processes are lacking, but substantial improvements in the global coverage of results is evident. Furthermore, even within those areas which have traditionally been the focus of catchment studies, new environments are being investigated. In this context, particular attention may be drawn to the establishments of studies in glacier basins as a means of assessing the contribution of glaciers to mechanical and chemical denudation. Work in Switzerland by Collins in the Gornera basin (Study 27) and by Gurnell (Study

Table 2.4. Suspended sediment and solute yields reported from selected catchment studies.

Study	Catchment and location	Mean annual sediment yield (t km^{-2} yr^{-1})
A. Suspended sediment yields		
73	Upper Cropp basin, New Zealand	29600 ± 2500
63	Tuanshangou basin, China	19751
61	Roma and Maliele catchments, Lesotho	220-1870
24	River Creedy catchment, UK	39
8	Piceance Creek, Colorado, USA	2.1-35.6
5	Watershed 10, H.J.Andrews Experimental Forest, Oregon	8.0
33	Starorobocianski stream, Poland	5.9
11	Hubbard Brook catchment, USA	3.3 (includes bed load)
B. Total dissolved solids loads		
41	Bystrzanka catchment, Poland	91.9
24	River Creedy catchment, JK	73.3
19	Merevale catchment, UK	46.8
33	Starorobocianski stream, Poland	39.3
5	Watershed 10, J.H.Andrews Experimental Forest, Oregon	33.0
8	Piceance Creek, Colorado, USA	13.2-24.9
11	Hubbard Brook catchment, USA	14.7

28) provide valuable results concerning the transport of particulate and dissolved material from the glacier snout and the processes operating in the proglacial zone.

Turning thirdly to a more quantitative aspect of the results obtained, it would seem justifiable to focus attention on the magnitude of sediment and solute yields documented in some of those studies where such measurements were undertaken, because these values provide a general indication of the rates of fluvial denudation operating in the landscape. These data are listed in Table 2.4, where particular emphasis has been given to high values in order to demonstrate the potential magnitude of contemporary denudation. Extremely high values of annual sediment yield in excess of 72102 t km^{-2}year^{-1} have been documented in the loess plateau region of the Yellow River basin in China by Mou and Meng (Study 63). Mean annual suspended sediment yields in this region may exceed 19000 t km^{-2}year^{-1} and, assuming a density of 1.5 t m^{-3} for the loess material, these values represent denudation rates in excess of 10 mm year^{-1}. A similarly high mean annual suspended sediment yield of 29600 t km^{-2}year^{-1} has been reported by Griffiths and McSaveney from their study of the Cropp River basin (Study 73). Annual rainfall in this mountain torrent basin in the western Southern Alps of South Island, New Zealand exceeds 10000 mm year^{-1} and a rainfall intensity of 680 mm in 24 hours has a return period of 2.3 years! Assuming a bedrock density of 2.65 t m^{-3}, these workers again point to denudation rates in excess of 10 mm year^{-1}.

Current estimates for global maximum levels of total dissolved solids yield point to values of approximately 500 t km^{-2} (cf. Walling and Webb, 1983). Values cited in Table 2.4 are considerably below this but, nevertheless, demonstrate considerable variation in loads between different catchments. These data cannot be used to provide estimates of chemical denudation rates, since a significant proportion of the load may be derived from non-denudational sources. In the case of both suspended sediment and total solute loads, the results of recent studies can be seen as providing improved knowledge of the range of values representative of different areas of the globe.

Interpretations of denudation rates based on measurements of material transport at catchment outlets are necessarily restricted by the lumped nature of the values obtained and by their failure to elucidate the relative importance of individual sediment sources and sediment generation processes and the significance of storage within the basin. The sediment budget concept advanced by Dietrich and Dunne (1978) provides a valuable means of evaluating the relative importance of various processes of sediment generation and storage and of interpreting the sediment yield from the basin in terms of the rates of operation of processes within the catchment. It is therefore suggested that the recent appearance of definitive data regarding the sediment and solute budgets of drainage basins could be highlighted as a fourth important achievement of current work.

Most of the work on sediment budgets has apparently been undertaken in the western United States and results obtained by Swanson and his colleagues for watershed 10 in the H.J. Andrews Experimental Forest in Oregon (Study 5) and by Lehre in the Lone Tree Creek basin in California (Study 7) are listed in Tables 2.5 and 2.6. In the case of the budget constructed by Swanson and his colleagues, information is provided concerning both the area of the watershed influenced by specific processes and their frequency

Table 2.5. Process characteristics and transfer rates of organic and inorganic material to the channel by hillslope processes (t $year^{-1}$) and export from the channel by channel processes (t $year^{-1}$) for Watershed 10, H.J.Andrews Experimental Forest, Oregon, USA.

Process	Frequency	Area influenced (%)	Material transfer, inorganic/organic	
Hillslope processes				
Solution transfer	Continuous	99	3	0.3
Litterfall	Continuous, seasonal	100	0	0.3
Surface erosion	Continuous	99	0.5	0.3
Creep	Seasonal	99	1.1	0.04
Root throw	1/year	0.1**	0.1	0.1
Debris avalanche	1/370 year	1-2**	6	0.4
Slump/earthflow	Seasonal*	58	0	0
Total			10.7	1.4
Channel processes				
Solution transfer	Continuous	1	3.0	0.3
Suspended sediment	Continuous, storm	1	0.7	0.1
Bedload	Storm	1	0.6	0.3
Debris torrent	1/580 year	1	4.6	0.3
Total			8.9	1.0

* Inactive in past century in Watershed 10.

** Area influenced by one event.

characteristics. This latter factor is extremely important in this basin since debris avalanches and torrents may only occur once in several centuries but nevertheless account for a large proportion of the long term sediment transfer. Comparison of the total transfers associated with hillslope and channel processes provides an indication of the importance of sediment storage within the basin.

The budget produced by Lehre and presented in Table 2.6 takes a slightly different form and provides additional detail on the proportion of the sediment mobilized by individual processes that reaches the channels. Furthermore, the possibility of comparing the budget estimates for individual years provides a means of evaluating the influence of the magnitude-frequency characteristics of the processes involved. The data indicate that

Table 2.6. Sediment mobilization and production by individual processes in the Lone Tree drainage basin.

Source	1971-1972		1972-1973	
	Mobilization (on slopes) (t km^{-2})	Production (reaching channels) (t km^{-2})	Mobilization (on slopes) (t km^{-2})	Production (reaching channels) (t km^{-2})
Landslides	0	0	1094	559
Slide scarp erosion	34	21	59	37
Slide scar sheet erosion	20	15	76	59
Headcut erosion	22	20	25	24
Soil creep	6*	Included in gully-	6*	Included in gully-
Hillslope sheet erosion	4*	scarp erosion	4*	scarp erosion
Gully-scarp erosion		82		194
Bank erosion		8		20
Bed erosion		2		92
Total	86	148	1219	985
	1973-1974		1972-1974 average	
Landslides	1795	1223	948	594
Slide scarp erosion	41	26	45	28
Slide scar sheet erosion	107	84	68	53
Headcut erosion	7	6	18	17
Soil creep	6*	Included in gully-	6*	Included in gully-
Hillslope sheet erosion	4*	scarp erosion	4*	scarp erosion
Gully-scarp erosion		201		159
Bank erosion		20		16
Bed erosion		15		37
Total	1960	1575	1089	904

*Indicates effective mobilization. Effective mobilization, computed as (total length of channel banks of drainage basin) % (mobilization rate per unit width of slope), gives rate at which sediment can be supplied to channel banks by processes acting continuously over drainage-basin hillslopes. In contrast, total mobilization (sediment moved by distance) by such processes is given by (thickness of moving layer) % (total area affected by process) and is independent of downslope transport velocity. Based on Lehre (1982).

Table 2.7. Annual uptake of nutrients by the vegetation of the Kali Mondo basin in relation to chemical denudation. Based on Bruijnzeel (1983).

	Ca	Mg	Na	K	SiO_2	P	Al	Fe	Mg
Uptake by vegetation ($kgha^{-1}year^{-1}$)	57.3	12.6	2.6	27.0	71.0	5.2	3.6	2.3	1.1
Apparent denudation ($kgha^{-1}year^{-1}$)	19.1	26.5	13.1	12.4	527	-0.2	14	13	1
Actual denudation ($kgha^{-1}year^{-1}$)	76.4	39.1	15.7	39.4	599	5	18	15	2
Apparent/actual (%)	24	67	83	30	88	-	82	85	50

in dry years, and in wet years without extreme flow events, most of the sediment mobilized goes into storage, chiefly on the lower parts of slopes and in channel and gully beds and banks. Large net removal of sediment from storage occurs in flow events with recurrence intervals greater than 10 to 15 years.

Detailed solute budgets have been produced for a number of studies listed in Table 2.1 (e.g. Studies 5, 10, 11, 14, 17, 19, 25, 30, 32, 49, 50, 52, 66 and 72) but in the majority of cases attention has focussed on the gross balance between precipitation inputs and outputs in streamflow. Johnson and Likens working in the Hubbard Brook catchment (Study 11) do, however, take account of nutrient uptake by vegetation. Bruijnzeel's study of the Kali Mondo basin in central Java (Study 66) is particularly important in this context in that he attempts to quantify the influence of nutrient uptake by the fast growing plantation forest on denudation rates. He calculates apparent denudation rates for a selection of solute constituents by comparing precipitation inputs with streamflow outputs, evaluates the magnitude of vegetation uptake, and provides estimates of actual denudation rates as the sum of the two former values (Table 2.7). In the extreme case of Ca, actual denudation is more than four times greater than the apparent denudation.

As a final general observation on the results of recent and ongoing catchment studies, reference may be made to a growing awareness of the need for an interdisciplinary approach and more particularly to the significance of biotic factors in influencing the operation of geomorphological processes. In this context, the work of Imeson and Jungerius and their colleagues in Luxembourg may be highlighted as demonstrating the need to consider the action of burrowing animals and earthworms in order to understand sediment production and movement on forested slopes. Similarly the work of Yair and his colleagues in the Sede Boker study in Israel has demonstrated how the pattern of sediment generation from limestone hillslopes in this area can only be explained if the activity of desert animals, especially isopods, is taken into account. The work of Gurnell and Gregory in the Highland Water catchment, UK (Study 22) also provides valuable evidence of the importance of vegetation composition in influencing runoff generation processes and of bank vegetation and debris dams in controlling channel development.

2.6.1 *Problems and the adequacy of available methods*

Looking generally at this sample of catchment studies of geomorphological processes, a number of important problems facing such investigations can be noted. The first relates to the costs involved in establishing detailed instrumentation networks, such as that depicted in Figure 2.9. Whilst it must be accepted that there is no direct relationship between the sum expended on instrumentation and the value of the results obtained, it is, nevertheless, clear that the products of new technology, including solid state data loggers, equipment for continuous monitoring of suspended sediment concentrations, and microprocessor controlled automatic water samplers, do offer considerable potential to the geomorphologist. Their cost is, however, usually high. A radio-telemetry system is doubtless beyond the scope of most, if not all, investigations but it would be difficult to deny that the potential for real-time data collection and therefore the possibility of planning visits to a field site to coincide with specific events would be of great value to many geomorphological studies.

In this context it is important that geomorphologists should be realistic and pragmatic about sources of funding and of equipment. Whereas elucidation of long-term landform development must remain a central goal of field investigations, this goal is essentially academic. Studies of contemporary processes may be more easily justified in terms of environmental management and other applied issues and such opportunities should be grasped. The work of Schick and his colleagues in the Nahel Yael basin in Israel (Study 57) provides an outstanding example of where the scope of a long-term geomorphological investigation has been extended to embrace such topics as flash flooding and stormwater management in order to ensure its continuation and to enhance its value.

It must also be recognized that many instruments, particularly modern electronic equipment, can be built by a skilled technician at a fraction of the cost of commercial equipment. Such purpose-built equipment may indeed be more closely suited to the job in hand. The recent initiative taken by a group of members of the British Geomorphological Research Group in commissioning the design of a solid state logger specifically adapted for geomorphological investigations and which could subsequently be built at cost by those involved in the scheme, can only be applauded. It must be accepted that a well-equipped mechanical and electronic workshop can be an important element of any intensive programme of catchment instrumentation.

A second problem is common to many field experiments and concerns the need to operate a measurement programme for a considerable period of time. There are certain studies where a short period of observation can, and indeed has, generated valuable results. This is the situation with many studies of runoff processes, where it is important to identify the mechanisms and pathways involved and where a longer period of observation is unlikely to produce any major additional benefits (e.g. Studies 15, 16, 20, 21). In many other instances, however, the value of the data base increases with the length of the period covered. This must be the case where an attempt is made to define the magnitude/frequency characteristics of fluvial transport (e.g. Webb and Walling, 1982b), or where a representative estimate of the long-term mean of a particular variable is required. For example, in some areas of the world the coefficient of variation (C_v) of a record of annual suspended sediment yield may approach 1.0. Simple standard error statistics indicate that in these circumstances a record of 100 years duration would be required in order to produce an estimate of the long-term mean to within an accuracy of ± 20% at the 95% level of confidence! More rigorous analysis involving appropriate statistical distributions could refine this estimate of the record length required, but it would not alter its order of magnitude. Furthermore, considerations of climatic change can only compound these problems.

The need for investigations of long duration inevitably introduces further problems in terms of cost or the provision of supporting staff and facilities over a considerable period of time, particularly within a financial environment which increasingly favours short term commitments. It is fortunate that some hydrological experiments such as those operated by forestry organizations are firmly committed to long-term monitoring and can therefore provide a basis for parallel geomorphological studies over the same timescale.

A final problem is one that is frequently cited in critiques of the role of representative and experimental basin studies in hydrology and relates to the transferability of results. Much of the ongoing work in catchment studies of geomorphological processes is essentially empirical in nature and is in many cases site-specific. There is a need for more effort in using the results obtained to understand the fundamental mechanisms and to develop a sound theoretical base which can in turn be used to assist extrapolation of results. Nevertheless, it must be accepted that the science of geomorphology is still in its infancy and empirical case studies are an essential prerequisite to many of the developments to which we may aspire in the future.

Turning more specifically to the methods employed in catchment studies, it would seem appropriate to highlight the problems associated with sampling. Spatial sampling inevitably assumes an important role in many studies, in view of the area associated with even a small drainage basin, and it could be suggested that insufficient attention is frequently paid to the associated problems. Transfer of attention from a catchment outlet

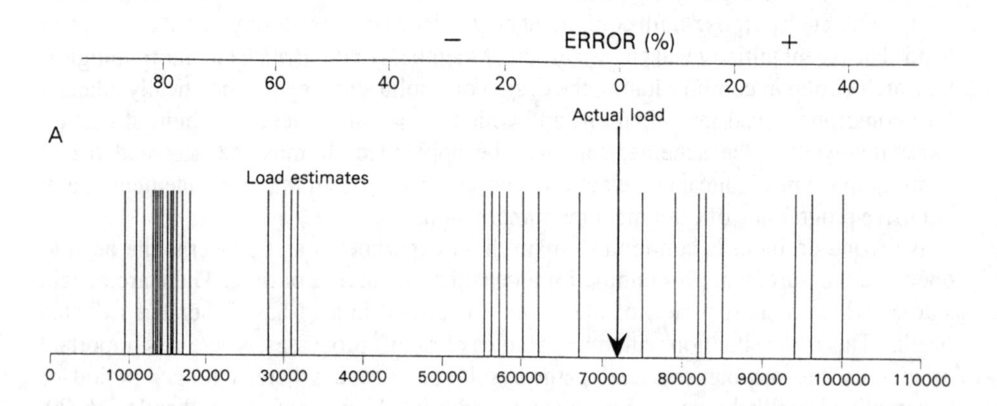

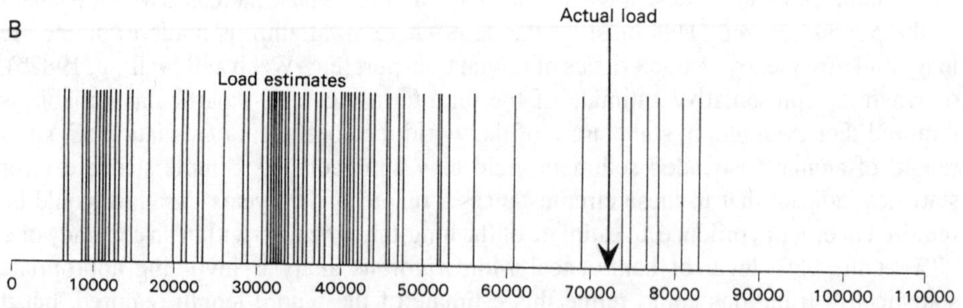

Figure 2.11. A comparison of suspended sediment load estimates for the River Creedy, obtained using interpolation and extropolation procedures, with the actual load for the period 1972-1979.

to processes operating within the basin must be accompanied by a recognition of its potential heterogeneity. Gallie and Slaymaker (Study 1) provide a useful example of the need to sample the entire range of soil-vegetation complexes within a basin in order to understand the interactions between solute generation and runoff pathways. Similarly the studies of runoff dynamics undertaken by Yair (Study 56), Anderson and Burt (Study 21) and Jones (Study 18) emphasise the considerable spatial variability of soil moisture conditions and runoff contributions that may occur over a hillslope. The geomorphologist can look to improved instrumentation and a denser network of measuring sites or to more rigorous structuring of his sampling strategy to minimise these problems. However, the latter may in practice depend heavily on the former to demonstrate the degree of variability involved and the influencing factors to be taken into account.

Problems associated with temporal sampling have been considerably reduced with the advent of continuous logging equipment but it is suggested that insufficient attention has been given to this problem in many studies concerned with measuring sediment and solute yields at catchment outlets. It is generally impossible to obtain a reliable estimate of the yield from infrequent sampling, since available interpolation and extrapolation procedures are inadequate to take account of the natural variability of fluvial transport. Figure 2.11, based on the work of Walling and Webb (Study 24), demonstrates the errors that may be involved in estimating suspended sediment yields from infrequent manual samples. It compares the suspended sediment load of the River Creedy, calculated using a continuous record of sediment concentration obtained from a turbidity monitor, with nearly onehundred load estimates for the same period derived using typical manual sampling strategies and a selection of load calculation procedures involving both interpolation and extrapolation (cf. Walling and Webb, 1981). The estimates span a wide range and underestimation by as much as 60% is common.

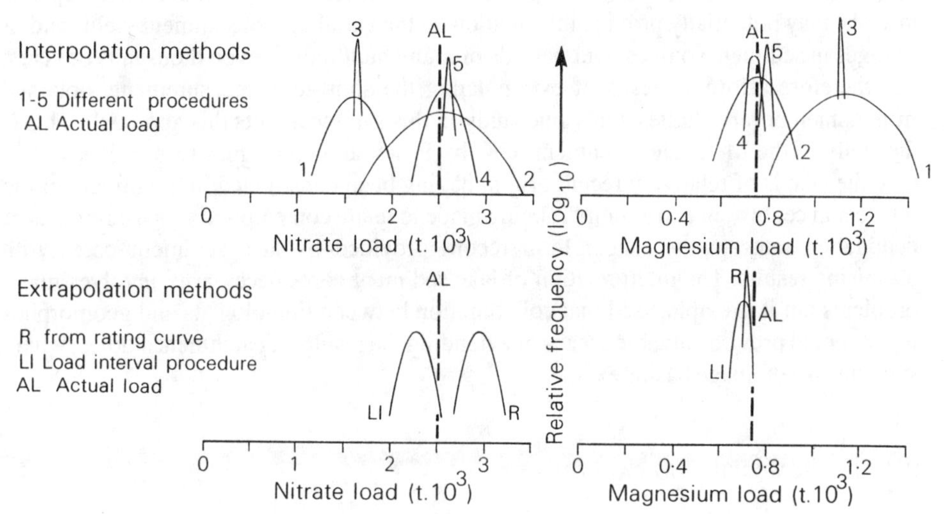

Figure 2.12. The accuracy and precision of estimates of nitrate and magnesium load for a 5 year period obtained using regular weekly sampling and various load calculation procedures.

Figure 2.12, based on studies by the same workers, compares accurate assessments of the nitrate and magnesium loads for a 5 year period for the River Dart with estimates based on regular weekly sampling and various load calculation procedures (cf. Walling and Webb, 1982). In this case both the accuracy and precision of the load estimates are represented. Replicate data sets have been generated for the specified sampling frequency and the variability of the resultant load estimates has been portrayed in an idealized form by plotting the normal distributions characterizing the associated values of mean and standard deviation. The ordinate scale has been logarithmically transformed and the distributions have been truncated at two standard deviations. The breadth of the distribution associated with a particular load calculation procedure therefore provides an indication of its precision, whereas the position of the distribution in relation to the actual load provides an indication of its accuracy. Errors as high as ± 40% and ± 25% could be associated with these estimates of nitrate and magnesium load respectively.

The problem of obtaining accurate estimates of river loads from infrequent samples has been aptly referred to as a 'geomorphic guessing game' by Olive et al. (1980). In most studies continuous records of suspended sediment concentrations and specific conductance or frequent samples collected using an automatic sampler will be required to produce reliable estimates of sediment and solute loads. Reliable estimates become essential when they are used to construct budgets since, for example, an underestimation of sediment yield in the development of a sediment budget could be erroneously interpreted as evidencing storage. Where techniques of questionable accuracy are employed to estimate loads, an indication of the potential errors should be provided in order to avoid such misinterpretation.

Returning finally to the more general problem of the length of time covered by a catchment study, this can be seen as a problem of temporal sampling. Even a period of 20 years may be insufficient to document the natural variability of the fluvial system in many environments, particularly where significant thresholds exist. There is no ready solution to this problem, but it is suggested that geomorphologists should give more attention to the potential of studying lake drainage basins (cf. Oldfield, 1977). The sediment deposits in a lake may potentially provide information on the variability of sediment yields and of changes in sediment sources over periods of many hundreds or even thousands of years, and therefore afford a means of extrapolating the short-term measurements obtained from contemporary studies. Only one study in this sample adopts this approach. This is the study of the Merevale catchment, UK, by Foster and colleagues (Study 19). In this case the lake is of relatively recent origin, having been constructed in the middle of the nineteenth century, but an attempt is being made to relate contemporary measurements of denudation rates to the longer term record provided by lake sediment cores with promising results. The interpretation of lake sediment cores necessarily involves many problems but it is emphasised that collaboration between limnologists and geomorphologists could prove a valuable means of extending the results of catchment studies, where suitable lake drainage basins exist.

2.7 THE GEOMORPHOLOGICAL CONTEXT

Perusal of the objectives of the various catchment studies included in this sample indicates that in nearly all cases the emphasis is on understanding the dynamics and monitoring the rate of operation of contemporary processes. Thus there are studies of the dynamics of runoff production on hillslopes and of sediment generation and conveyance and investigations of the magnitude of the loads transported by particular rivers. This approach is essentially that of 'functional' geomorphology as defined by Ahnert (1980) and in which he sees the study of process response systems as the major goal. It is not proposed to debate the issue of what should constitute the primary goal of geomorphological studies, but it can be suggested that a large number of the workers represented in this sample of studies are able to justify their work in terms of providing an improved understanding of contemporary processes, without recourse to considerations of long-term evolution of landforms and relief development. Several studies concerned with measuring river loads do produce estimates of denudation rates, but these are again used to index current rates of landscape development rather than to explain landform development.

Reasons for this emphasis on 'functional' geomorphology may perhaps be found in its relevance to significant environmental issues such as land degradation and, therefore, in the environmental manager directing or supporting research 'calling the tune'. The catchment is frequently employed as the fundamental unit for environmental management and its dynamics must be understood. To take one example, recent concern for the movement of contaminants through terrestrial ecosystems has highlighted the importance of sediment-associated transport and the significance of stores and sinks within the fluvial system. It is the geomorphologist who is probably best qualified to improve our understanding of these sinks and stores.

There is, however, no clear boundary between studies of contemporary processes and those of relief development and longer-term landform evolution. Reaction to the Davisian approach could be seen as having pushed interest towards the former, but there is clearly a renewal of interest in the latter. Furthermore, the former must itself provide the basis for future advances in our comprehension of the latter. It is thus encouraging to note that the results of studies of solutional denudation on hillslopes using micro-weight loss tablets are currently being used to test models of slope development (Studies 15 and 21) and that the objectives of Imeson and Jungerius and their co-workers in the Luxembourg Ardennes (Study 42) include studies of cuesta development (cf. Jungerius and Van Zon, 1982). If the development of models of contemporary geomorphic processes is seen as a central objective of current studies, improvement and calibration of these models should lead directly to potential for their extension through time and therefore for providing a basis for modelling long-term relief.

ACKNOWLEDGEMENT

The author is indebted to the many people who generously provided information on their catchment investigations for use in this review.

REFERENCES

Afifi, A.A. and Brickner, O.P., 1983. Weathering reactions, water chemistry and denudation rates in drainage basins of different bedrock types: 1 - sandstone and shale. *Dissolved loads of rivers and surface water quantity / quality relationships*. IAHS Publication No. 141, 193-203.

Ahnert, F., 1980. A note on measurements and experiments in geomorphology. *Z. Geomorph., Supp. Bd.* 35, 1-10.

Anderson, M.G. and Burt, T.P., 1977. Automatic monitoring of soil moisture conditions in a hillslope spur and hollow. *J. Hydrology*, 33, 27-36.

Anderson, M.G. and Burt, T.P., 1978. The role of topography in controlling throughflow generation. *Earth Surface Processes* 3, 331-344.

Anderson, M.G. and Kneale, P.E., 1982. The influence of low-angled topography on hillslope soil-water convergence and stream discharge. *J. Hydrology*, 57, 65-80.

Andersson-Calles, U.M. and Eriksson, E., 1979. Mass balance of dissolved inorganic substances in three representative basins in Sweden. *Nordic Hydrology*, 1979, 99-114.

Andrews, E.D., 1983. Denudation of the Piceance Creek basin, Colorado. *Dissolved loads of rivers and surface water quantity/quality relationships*, IAHS Publication 141, 205-215.

Association Geographique D'Alsace, 1982. Structure et fonctionnement du milieu naturel en moyenne montagne. *Recherches Geographiques à Strasbourg*, 19, 20 and 21.

Balteanu, D., Mihaiu, G., Negut, N. and Caplescu, L., 1984. Sources of sediment and channel changes in small catchments of Romania's hilly regions. In Burt, T.P. and Walling, D.E. (eds.), *Catchment experiments in fluvial geomorphology*, 277-288. Geobooks, Norwich, UK.

Becchi, I., Billi, P. and Tacconi, P., 1979. Field research on sediment production in small basins with different land use. *The hydrology of areas of low precipitation*, IAHS Publication 128, 409-420.

Bonell, M., Cassells, D.S. and Gilmour, D.A., 1982. Vertical and lateral soil water movement in a tropical rainforest catchment. *National symposium on forest hydrology*, Institute Engineers Australia, Publication 82/6.

Bruijnzeel, L.A., 1983. The chemical mass balance of a small basin in a wet monsoonal environment and the effect of fast-growing plantation forest. *Dissolved loads of rivers and surface water quantity/quality relationships*, IAHS Publication 229-239.

Bryan, R.B. and Campbell, I.A., 1982. Surface flow and erosional processes in semiarid mesoscale channels and drainage basins. *Recent developments in the explanation and prediction of erosion and sediment yield*, IAHS Publication 137, 123-133.

Bryan, R.B. and Hodges, W.K., 1984. Runoff and sediment transport dynamics in Canadian badland microcatchments. In Burt, T.P. and Walling, D.E., (eds.), *Catchment experiments in fluvial geomorphology*, 115-132. Geobooks, Norwich, UK.

Burt, T.P., Butcher, D.P., Coles, N. and Thomas, A.D., 1983. The natural history of Slapton Ley Nature Reserve XV: Hydrological processes in the Slapton Wood catchment. *Field Studies*, 5, 731-752.

Burt, T.P. and Gardiner, A.T., 1984. Runoff and sediment production in a small peat-covered catchment: Some preliminary results. In Burt, T.P. and Walling, D.E. (eds.), *Catchment experiments in fluvial geomorphology*, 133-152. Geobooks, Norwich, UK.

Burt, T.P. and Walling, D.E., eds., 1984. *Catchment experiments in fluvial geomorphology*. Geobooks, Norwich, UK.

Chakela, Q.K., 1981. *Soil erosion and reservoir sedimentation in Lesotho*. Uppsala Universitat Naturgeografiska Institutionen Rapport No. 54.

Church, M., 1981. On experimental method in geomorphology. In Burt, T.P. and Walling, D.E. (eds.), *Catchment experiments in fluvial geomorphology*, 563-580. Geobooks, Norwich, UK.

Collins, D.N., 1983. Solute yield from a glacierized high mountain basin. *Dissolved loads of rivers and surface water quantity/quality relationships*, IAHS Publication 141, 41-49.

Decoursey, D.G., 1982a. The Goodwin Creek Research Catchment Part I. Design Philosophy. *Hydrological research basins and their use in water resources planning*, Landeshydrologie, Berne, 65-73.

Decoursey, D.G., 1982b. ARS's small watershed model. *Amer. Soc. Agric. Engineers. Summer Meeting Madison Wisconsin*, June 1982, Paper 82.2094.

Dietrich, W.E. and Dunne, T., 1978. Sediment budget for a small catchment in mountainous terrain. *Z. Geomorph. Supp. Bd.* 29, 191-206.

Dupraz, C., Lelong, F., Troy, J.P. and Dumazet, B., 1982. Comparative study of the effects of vegetation on the hydrological and hydrochemical flows in three minor catchments of Mont Lozere (France) – Methodological aspects and first results. *Hydrological research basins and their use in water resources planning*, Landeshydrologie, Berne, 671-681.

Emmett, W.W., 1981. Measurement of bed load in rivers. *Erosion and sediment transport measurement*, IAHS Publication 133, 3-15.

Fish, I.L., 1983. Partech turbidity monitors: calibration with silt and the effects of sand. *Technical note OD/TN1, hydraulics research*, Wallingford, UK.

Foster, I.D.L., Carter, A.D. and Grieve, I.C., 1983. Biogeochemical controls on river water quality in a forested drainage basin, Warwickshire, UK. *Dissolved loads of rivers and surface water quantity/quality relationships*. IAHS Publication 141, 241-253.

Froehlich, W., 1983. The mechanisms of dissolved solids transport in flysch drainage basins. *Dissolved loads of rivers and surface water quantity/quality relationships*. IAHS Publication 141, 99-108.

Gallie, T.M. and Slaymaker, H.O., 1984. Variable solute sources and hydrologic pathways in a coastal subalpine environment. In Burt, T.P. and Walling, D.E. (eds.), *Catchment experiments in fluvial geomorphology*, 347-358. Geobooks, Norwich, UK.

Gladki, H. and Czaszynski, A., 1981. Simulation of the bed load transport in a mountain stream. *Proc. international conference on numerical modelling of river channel and overland flow for water resources and environmental applications*, Bratislava, May 4-8, 1981.

Griffiths, G.A. and Hicks, D.M., 1980. Transport of sediment in mountain streams: performance of a measurement system during a two year storm. *J. Hydrology* (NZ) 19, 131-136.

Gurnell, A.M., 1982. The dynamics of suspended sediment concentration in an alpine proglacial stream network. *Hydrological aspects of alpine and high mountain areas*, IAHS Publication 138, 319-330.

Gurnell, A.M. and Gregory, K.J., 1984. The influence of vegetation on stream channel processes. In Burt, T.P. and Walling, D.E. (eds.), *Catchment experiments in fluvial geomorphology*, 515-536. Geobooks, Norwich, UK.

Hasholt, B., 1976. Hydrology and transport of material in the Sermilik area 1972. *Geografisk Tidsskrift* 75, 30-39.

Hayward, J.A., 1980. Hydrology and stream sediment from Torlesse stream catchment. Tussock Grasslands and Mountain Lands Institute, Lincoln College, Special Publication No. 17.

Imeson, A.C. and Van Zon, H.J.M., 1979. Erosion processes in small forested catchments in Luxembourg. In Pitty, A.F. (ed.), *Geographical approaches to fluvial processes*, Geobooks, Norwich, UK, 93-107.

Jeje, L.K. and Nabega, A., 1983. Sediment yield in response to rainstorms and landuse in small basins in the Ile-Ife area of S.W. Nigeria. *Nigerian Geog*. J. 25.

Jones, J.A.A. and Crane, F.G., 1984. Pipeflow and pipe erosion in the Maesnant experimental catchment. In Burt, T.P. and Walling, D.E. (eds.), *Catchment experiments in fluvial geomorphology*, 55-72. Geobooks, Norwich, UK.

Jungerius, P.D. and Van Zon, H.J.M., 1982. The formation of the Lias Cuesta (Luxembourg) in the light of present-day erosion processes operating on forest soils. *Geografiska Annaler* 64A, 127-140.

Kang Zhicheng and Zhang Shucheng, 1984. An analysis of sediment transport by debris flows in the Jiangjia gully. In Burt, T.P. and Walling, D.E. (eds.), *Catchment experiments in fluvial geomorphology*, 477-488. Geobooks, Norwich, UK.

Keller, H.M. and Strobel, T., 1982. Water balance and nutrient budgets in subalpine basins of different forest cover. *Hydrological research basins and their use in water resources planning*, Landeshydrologie, Berne, 683-694.

Kirkby, M.J. and Chorley, R.J., 1967. Throughflow, overland flow and erosion. *Bull. International Assoc. Scientific Hydrology* 12, 5-21.

Knisel, W.G., ed., 1980. CREAMS: A field-scale model for chemicals, runoff and erosion from agricultural management systems. *US Dept. Agric. Conservation Research Report*, No. 26.

Krzemien, K., 1982. Removal of dissolved material from the Starorobocianski stream crystalline catchment basin (the Western Tatra Mts.). *Studia Geomorphologica Carpatho-Balcanica* 15, 70-80.

Le Roux, J.S. and Roos, Z.N., 1979. Rate of erosion in the catchment of the Bulbergfontein Dam near Reddersburg in the Orange Free State. *J. Limnological Society South Africa* 5, 89-93.

Le Roux, J.S. and Roos, Z.N., 1982. The rate of soil erosion in the Wuras Dam catchment calculated from sediments trapped in the dam. *Z. Geomorph.* 26, 315-329.

Lehre, A.K., 1982. Sediment budget of a small Coast Range drainage basin in North-Central California. *Sediment budgets and routing in forested drainage basins.* US Forest Service General Technical Report PNW-141, 67-77.

Lekach, J. and Schick, A.P., 1982. Suspended sediment in desert floods in small catchments. *Israel J. Earth-Sciences* 31, 144-156.

Leopold, L.B., 1962. The Vigil Network. *Bull. International Assoc. Scientific Hydrology* 7, 5-9.

Leopold, L.B. and Emmett, W.W., 1965. Vigil Network sites: a sample of data for permanent filing. *Bull. International Assoc. Scientific Hydrology* 10, 12-21.

Likens, G.E., Bormann, F.H., Pierce, R.S., Eaton, J.S. and Johnson, N.M., 1977. *Biogeochemistry of a forested ecosystem*, Springer Verlag, New York.

Lougran, R.J., Campbell, B.L. and Elliott, G.L., 1982. The identification and quantification of sediment sources using ^{137}Cs. *Recent developments in the explanation and prediction of erosion and sediment yield*, IAHS Publication 137, 361-369.

Madeyski, M., 1982. An expression for suspended load transportation due to high discharge in small Flysch river basins. *Studia Geomorphologica Carpatho-Balcanica*, 16.

Martin, C., 1981. L' établissement de bilans hydrogéochimiques dans la partie occidentale du Massif des Maures. Approche méthodologique. *Physio-Geo* 2, 39-58.

Mou Jinze and Meng Qingmei, 1982. Sediment delivery ratio as used in the computation of watershed sediment yield. *J. Hydrology* (NZ), 20, 27-38.

Newson, M.D., 1980. The geomorphological effectiveness of floods – a contribution stimulated by two recent events in mid-Wales. *Earth Surface Processes* 5, 1-16.

Nortcliff, S. and Thornes, J.B., 1981. Seasonal variations in the hydrology of a small forested catchment near Manaus, Amazonas, and the implications for its management. In Lal, R. and Russell, E.W. (eds.), *Tropical agricultural hydrology*, Wiley, Chichester, 37-57.

O'Loughlin, C.L., Rowe, L.K. and Pearce, A.J., 1978. Sediment yields from small forested catchments, North Westland-Nelson, New Zealand. *J. Hydrology* (NZ), 17, 1-11.

O'Loughlin, C.L., Rowe, L.K. and Pearce, A.J., 1980. Sediment yield and water quality responses to clearfelling of evergreen mixed forests in western New Zealand. *The influence of man on the hydrological regime with special reference to representative and experimental basins IAHS* Publication 130, 285-292.

O'Loughlin, C.L., Rowe, L.K. and Pearce, A.J., 1982. Exceptional storm influences on slope erosion and sediment yield in small forest catchments, north Westland, New Zealand. *National symposium on forest hydrology, Institute of Engineers Australia* Publication 82/6, 84-91.

Oldfield, F., 1977. Lakes and their drainage basins as units of sediment based ecological study. *Progress in Physical Geog.* 1, 460-504.

Olive, L.J., Rieger, W.A. and Burgess, J.S., 1980. Estimation of sediment yields in small catchments: a geomorphic guessing game? *Proc. Conference Institute Australian Geographers*, Newcastle, NSW.

Owens, L.B. and Watson, J.P., 1979. Landscape reduction by weathering in small Rhodesian watersheds. *Geol.* 7, 281-84.

Parizek, R.R. and Lane, B.E., 1970. Soilwater sampling using pan and deep pressure vacuum lysimeters. *J. Hydrology*, 11, 1-21.

Peart, M.R. and Walling, D.E., 1982. Particle size characteristics of fluvial suspended sediment. *Recent developments in the explanation and prediction of erosion and sediment yield.* IAHS Publication 137, 397-407.

Reid, J.M., Macleod, D.A. and Cresser, M.S., 1981. The assessment of chemical weathering rates within an upland catchment in North-East Scotland. *Earth Surface Processes and Landforms* 6, 447-457.

Rieger, W.A., Olive, L.J. and Burgess, J.S., 1982. The behaviour of sediment concentrations and solute concentrations in small forested catchments. *National symposium on forest hydrology, Institute Engineers Australia* Publication 82/6, 79-83.

Ritchie, J.C., Spraberry, J.A. and McHenry, J.R., 1974. Estimating soil erosion from the redistribution of fallout Cs-137. *Soil Science Soc. Amer Proc.* 38, 137-139.

Roberts, G., Hudson, J.A. and Blackie, J.R., 1983. Nutrient cycling in the Wye and Severn at Plynlimon. *Institute of hydrology report* 86, Wallingford, UK.

Rodda, J.C., 1976. Basin studies. In Rodda, J.C. (ed.), *Facets of hydrology* Wiley, Chichester, 257-297.

Sala, M., 1981. Geomorphic processes in a small Mediterranean drainage basin (Catalan Ranges). *Trans. Japanese Geomorphological Union* 2, 239-252.

Slaymaker, O., 1980. Geomorphic field experiments. Inventory and prospect. *Z. Geomorph. Supp.* Bd. 35, 183-194.

Slaymaker, O., Dunne, T. and Rapp, A., 1980. Preface to Geomorphic Experiments on Hillslopes. *Z. Geomorph. Supp. Bd.* 35: V-VII.

Smart, C.C. and Ford, D.C., 1982. Quantitative dye tracing in a glacerised alpine karst. *Beitraege zur geologie der Schweiz – Hydrologie* 28, 191-200.

Smart, P.L. and Laidlaw, I.M.S., 1977. An evaluation of some fluorescent dyes for water tracing. *Water Resources Research* 13, 15-33.

Swanson, F.J., Fredriksen, R.L. and McCorison, F.M., 1982. Material transfer in a western Oregon Forested Watershed. In Edmonds, R.L. (ed.), *Analysis of coniferous forest ecosystems in the Western United States*, US/IBP Synthesis Series 14. Hutchinson Ross, Stroudsburg, PA, 233-266.

Tanaka, T., 1982. The role of subsurface water exfiltration in soil erosion processes. *Recent developments in the explanation and prediction of erosion and sediment yield,* IAHS Publication 137, 73-80.

Tazioli, G.S., 1981. Nuclear techniques for measuring sediment transport in natural streams – examples from instrumented basins. *Erosion and sediment transport measurement*, IAHS Publication 133, 63-81.

Trudgill, S.T., Crabtree, R.W., Pickles, A.M., Smettem, K.R.J. and Burt, T.P. Hydrology and solute uptake in hillslope soils on magnesian limestone. In Burt, T.P. and Walling, D.E. (eds.), *Catchment experiments in fluvial geomorphology*, 183-215. Geobooks, Norwich, UK.

Walling, D.E., 1983. The sediment delivery problem. *J. Hydrology*, 69, 209-237.

Walling, D.E. and Webb, B.W., 1981. The reliability of suspended sediment load data. *Erosion and sediment transport measurement*, IAHS Publication 133, 177-194.

Walling, D.E. and Webb, B.W., 1982. The design of sampling programmes for studying catchment nutrient dynamics. *Hydrological research basins and their use in water resources planning.* Landeshydrologie, Berne, 747-758.

Walling, D.E. and Webb, B.W., 1983. The dissolved loads of rivers: a global overview. *Dissolved loads of rivers and surface water quantity/quality relationships*, IAHS Publication 141, 3-20.

Webb, B.W. and Walling, D.E., 1982. The magnitude and frequency characteristics of fluvial transport in a Devon drainage basin and some geomorphological implications. *Catena* 9, 9-23.

Webb, B.W. and Walling, D.E., 1982. Catchment scale and the interpretation of water quality behaviour. *Hydrologic research basins and their use in water resources planning.* Landeshydrologie, Berne, 759-770.

Webb, B.W. and Walling, D.E., 1983. Stream solute behaviour in the River Exe basin, Devon, UK. *Dissolved loads of rivers and surface water quantity/quality relationships.* IAHS Publication 141, 153-169.

Welc, A., 1978. Spatial differentiation of chemical denudation in the Bystrzanka flysch catchment. *Studia Geomorphologica Carpatho-Balcanica* 12, 149-162.

Williams, A.G., Ternan, J.L. and Kent, M., 1983. Stream solute sources and variations in a temperate granite drainage basin. *Dissolved loads of rivers and surface water quantity/quality relationships.* IAHS Publication 141, 299-310.

Yair, A., 1981. The Sede Boker experiment site. In Dan, J., Gerson, R., Koyumojisky, H. and Yaalon, D.H. (eds.), *Aridic soils of Israel*, Agric. Research Organization Institute of Soils and Water Special Publication 190, 239-254.

Slaymaker, O., Sanderson, [illegible] and McPherson, [illegible] 19[illegible] Estimating soil erosion from the [illegible] Fraser [illegible] Soil Science Society [illegible] 36, [illegible]

[illegible], G., Jackson, [illegible] Harris, [illegible] 1983 [illegible] in the [illegible] *Fluvial [illegible]* [illegible], Wallingford, UK [illegible]

[illegible] 1[illegible] *Using [illegible] in Hydrology* [illegible]

[illegible], M., 1985 [illegible] processes in a small [illegible] drainage basin [illegible] *Earth Surface Processes and Landforms* [illegible]

[illegible], D. 1981 [illegible] *Geografiska Annaler* [illegible] 63A, [illegible]

[illegible] 1[illegible] *Geomorphology* [illegible]

[illegible] and [illegible] 1982 [illegible] *Geografiska Annaler* 64, [illegible]

[illegible] 1[illegible] *Zeitschrift [illegible]* [illegible] 1[illegible]

Swanson, F.J., [illegible] and [illegible] 1982 [illegible] in a forested drainage basin [illegible] Swanson, F.J., [illegible] *Sediment Budgets and Routing in Forested Drainage Basins* [illegible] General Technical Report PNW-141 [illegible]

[illegible] 19[illegible] The role of [illegible] *Sediment Budgets* IAHS Publication [illegible]

[illegible] 19[illegible] [illegible] *Sediment Budgets* [illegible] IAHS Publication 1[illegible]

[illegible], Crowder, [illegible] and [illegible] 19[illegible] [illegible] and [illegible] relationships [illegible]

Walling, D.E. 1983 The sediment delivery problem. *Journal of Hydrology* 65, 209-37

Walling, D.E. and Webb, B.W. 1983 The [illegible] sediment and [illegible] IAHS Publication [illegible]

Walling, D.E. and Webb, B.W. 1982 [illegible] approaches to studying [illegible] in [illegible] *Earth Surface Processes and Landforms* [illegible]

Walling, D.E. and Webb, B.W. 1987 [illegible] IAHS Publication 1[illegible]

Webb, B.W. and Walling, D.E. 1982 [illegible] *Catena* 9, [illegible]

Webb, B.W. and Walling, D.E. 198[illegible] [illegible] *Journal of Hydrology* [illegible]

Webb, B.W. and Walling, D.E. 198[illegible] [illegible] IAHS Publication [illegible]

Wilkinson, [illegible] [illegible] *[illegible] Landforms* [illegible]

Williams, R.G., [illegible] and [illegible] 1[illegible] [illegible] and variations in a temperate [illegible] IAHS Publication 1[illegible]

[illegible] 19[illegible] [illegible] Research Council [illegible] Publication [illegible]

CHAPTER 3

Rapid mass movement

SETSUO OKUDA
Okayama University of Sciences, Japan

3.1 INTRODUCTION

3.1.1 *Rapid mass movement and field experiments*

The term 'mass movement' means a geomorphological process that involves a transfer of slope materials downwards under the influence of gravity without the primary assistance of other transporting agents – wind, water, or ice. The terminological distinction between 'rapid' and 'slow' movement is a qualitative one. 'Rapid' mass movement is interpreted as mass movement with a speed which can be recognized by human sense without any special measuring instruments.

Rapid mass movement brings about observable geomorphological change during periods shorter than a human life span. It is of interest to students of contemporary geomorphic process as well as to students of longer term slope evolution. Apart from the academic view point, rapid mass movement is an object of disaster prevention science in many countries. It is often responsible for various types of natural hazard through rapid erosion and/or sediment accumulation in residential areas.

To promote the scientific study of rapid mass movement, various approaches are necessary, for example, field and/or laboratory experiments, numerical analysis and model construction. The author believes that field experimentation is a particularly effective approach in the study of rapid mass movement, because the actual geomorphological changes can be observed directly and the relationship of their changes to exogenetic stresses can be established more easily and quantitatively than in the case of many other geomorphological problems.

In fact, many field experiments on rapid mass movement are now being carried on in various countries as shown in Figure 3.1 which was the result of questionnaires (Table 3.1). In this report, some field experimental methods are introduced which have been applied in the fields listed in Table 3.1 and found useful for the study of rapid mass wasting. The major themes which are addressed are: (a) slope morphometry, (b) in situ measurement, (c) hydrology (d) dynamic phenomena, (e) laboratory work, modelling and simulation, and (f) hillslope evolution.

Table 3.1. Summary of rapid mass movement experiments.

No.	Country	Researcher (answerer)	Category I						Category II						Research area (period)
			a	b	c	d	e	f	g	h	i	j	k	l	
76	Canada	Bovis, M.J.		x							x	x			Coast Mts., British Columbia (1980-)
77	Canada	Gardner, J.	x								x	x			Rocky Mts., Alberta (1972-)
78	Canada	Luckman, B.	x								x	x			Rocky Mts., BC/Alberta (1972-)
79	USA	O'Brien, J.					x		x	x	x	x	x		Steep basin around Glenwood Springs, Col.
80	USA	Onesti, L.J.				x	x		x	x	x				Atigun Pass Region, Central Brooks Range Alaska (1978-)
81	USA	Dunne, T.					x		x	x			x		Toutle Valley, Mt. St. Helens
82	USA	Kelsey, H.M.				x	x					x			Van Duzen Basin, California
83	UK	Brunsden, D.						x		x	x				Nepal, Pakistan Algeria, UK
84	UK	Brunsden, D.			x						x	x	x		West Dorset Coast (1966-)
85	UK	Brunsden, D.					x							x	Karakoram Mt. Pakistan
86	UK	Chandler, R.J.					x				x				Man-made clay slopes in England
87	UK	Hutchinson, J.N.	x	x	x	x	x				x	x	x	x	Not specified
88	UK	Kirkby, M.J.			x									x	Pendine-Laugharne, S. Wales
89	UK	Whalley, W.B.	x		x		x				x	x	x	x	Minis North, Co. Antrim, N. Ireland (1976-), Iceland (1979-)
90	Sweden	Nyberg, R.						x			x				Abisko-Kebnekaise mountains, Swedish Lappland
91	Poland	Jahn, A.			x				x			x			Stolowe Gory Mts., South Poland (1973-)
92	Poland	Kotarba, A.	x		x		x			x		x			High Tatra Mt.
93	Germany	Grunnert, J.			x									x	Northern Africa, especially the escarpments of Murzuk-Basin
94	Germany	Richter, G.			x	x	x								Vineyard in Moselle Valley (1974-)
95	Germany	Abele, G.													
96	Netherlands	Van Asch, Th.W.J.			x	x	x				x	x	x	x	Landslide area near La Mure (1979-82)
97	Netherlands	Van Asch, Th.W.J.					x	x							Lower Rhone Valley, French Alps
98	Italy	Govi, M.	x		x		x			x	x				Po river basin and other areas (1970-)
99	Italy	Edoardo, S.													Alpapo, Bellumo Province, Italy (1966-)
100	Italy	Canuti	x		x	x	x	x		x	x	x			Tuscany (1978-)
101	Romania	Balteanu, D.			x	x			x	x		x			Buzau Subcarpathians (1968-)
102	Romania	Balteanu, D.					x	x	x	x		x		x	Buzau Subcarpathians (1975-)
103	Romania	Surdeanu, V.			x	x	x		x	x	x				Pingarati Carpathians (1975-)
104	Japan	Ashida, K.	x		x		x				x		x		Ashiarai Valley, Western Slope of Mt. Yake (1967-)

Field work method	Laboratory work method	Model or simulation	Most recent reference paper
Surveying and geophysical	geotechnical testing		(1982)
Measurement of movement	physical properties		(1980)
Measurement of movement	physical properties		(1982)
Investigation of debris and mudflow	property of mudflow deposits	numerical model	(1982)
Meteorological and snow observation, analysis of deposits	grain size	numerical model	(1980)
Survey of evidence of flow mechanics and sediment transport	flume experiments	analytical model	(1982)
Observation of earth flow movement			(1978)
Geomorphological mapping, subsurface properties	geotechnical testing	stability analysis	(1980)
Surveying, auguring, seismic techniques	geotechnical testing	static model for stability analysis, dynamic model during motion	(1979)
Geomorphological mapping, surveying	geotechnical testing	rheological models	in progress for report
Boreholes, porepressure, movement of landslide			(1974)
Geomorphological surveys and geophysical exploration	geotechnical tests	slope stability analysis	(1982)
Use of existing data		computer simulation	in preparation
Measurement of movement, pore water pressure	scale model experiments of slides		
Material transport, property of snow			(1982)
Three dimensional measurement of block sliding			(1977)
Geomorphological mapping, terrestrial photo, measurement of movement			(1981)
Cartography, geological profiles	clay mineral and thin-section analysis		(1980)
Geomorphological and soil mapping	granulometry, water-content		(1982)
No field experiments			some related papers
3eodetic survey, pore pressure, sliding feature	geotechnical tests	developing viscous models	
Sedimentological survey	flume experiments, soil mechanical analyses		in preparation
Synthetic	various kinds		(1982)
3eological survey, movement measurement	clay analysis		(1982)
3round survey	geotechnical testing		(1979)
'hotogrammetrical; stakes	geotechnical testing		(1984)
'hotogrammetrical	geotechnical testing		(1984)
Surface & subsurface flow	geotechnical testing	slope stability analysis	(1984)
Synthetic observation of debris flows			(1980)

Table 3.1 (continued).

No.	Country	Researcher (answerer)	Category I a	b	c	d	e	f	Category II g	h	i	j	k	l	Research area (period)
105	Japan	Fujita, T.			x	x					x				Land creep field, Shikoku and Sanda basin in Kobe City
106	Japan	Furuya, T.			x	x				x	x	x			Land creep field, Mineoka Mts. Chiba Pref. (1979-)
107	Japan	Ishii, T.	x										x		Ashio Mts., Kanto District (1982-)
108	Japan	Kadomura, H.						x		x	x	x			Usu Vocano, Southwest Hokkaido (1977-)
109	Japan	Mizuyama, T.			x	x	x			x	x		x		Sakurajima (Kyushu), Mt. Yake (Central Japan, Mt. Fuji (Central Japan), Mt. Usu (Hokkaido)
110	Japan	Okimura, T.			x		x			x	x		x		Aotani, Rokko Mts. Kobe (1974-)
111	Japan	Okuda, S.	a x				x			x	x	x			Mt. Yakedake, Nagano, Pref. Central Japan (1970-)
112	Japan	Okunishi, K.			x				x	x	x	x	x		Nishimikawa Mts. Aichi Pref. (1975-)
113	Japan	Oyagi, N.			x			x			x				Ohike, Shizuoka Pref.
114	Japan	Oyagi, N.								x					National Research Center of Disaster Prevention
115	Japan	Sassa, K.			x		x		x		x				Land creeping fields in Japan
116	Japan	Tsukamoto, Y.			x						x				University forest for field test, Tokyo
117	PRC	Ding, X.G.			x	x				x	x				
118	PRC	Ding, X.G.					x								Jiangjia Ravine, Yunnan Prov. (1965-)
119	New Zeal.	Crozier, M.J.	x	x	x	x	x	x			x				New Zealand (1974-)
120	New Zeal.	Crozier, M.J.			x				x						Wairarapa (1977-1983)
121	New Zeal.	Pearce, A.J.			x				x	x	x	x	x		Norown, Westland, South Island (1973-79)
122	New Zeal.	Pearce, A.J.			x		x		x	x	x	x	x		Maimai, North Westland, South Island (1973)
123	New Zeal.	Pearce, A.J.					x					x	x		Mt. Thomas, North Canterbury, South Island (1978-1980)
124	New Zeal.	Selby, M.J.			x				x		x				Hillcountry, North Island, New Zealand (1978-)
125	New Zeal.	Vincent, K.W.			x					x					Te Whanga Station Wairarapa, New Zealand (1980-83)

a = falls, b = topples, c = slides, d = slumps, e = flows, f = various, g = processes causing r.m.w., h = r.m.w. and hazard damage, i = prediction and/or control of r.m.w., j = topographic change and r.m.w., k = physical modelling, l = other.

Field work method	Laboratory work method	Model or simulation	Most recent reference paper
Geophysical exploration, geological survey	geotechnical tests	slope stability	(1981)
Observation of slope deformation	airphoto analysis		(1983)
Observation of talus stability	model experiments of falling stone		(1978)
Slope deformation, motion of debris flow	air photo analysis, subsurface property, etc.		(1983)
Synthetic observation of debris flows	flume experiments of debris flows	physical or stochastic model of debris flow	(1983)
Synthetic observation of slope hydrology	airphoto analysis, soil test	simulation of groundwater flow	(1983)
Synthetic observation on debris flow	grain size, inverse sorting		(1982)
Physical property of top soil, hydrological observation	measurement of permeability of samples	physical modelling	(1983)
Trench cut, sampling, slide profile	analysis of samples		(1983)
Slope movement, soil and water pressure	model experiment of slope failure	hydraulic experiment	(1978)
Synthetic observation on land creep	geotechnical tests, mechanism of debris flows initiation		(1982)
Measurement of subsurface flow		computer simulation of subsurface flow	(1983)
	hydraulic model exp. on debris flow		in progress for report
Observation on mud flow	hydraulic model exp. on debris flow		(1981)
Observation on movement of soil	soil moisture capacity		(1981)
Electronic data logging	physical property of soil		(1982)
Landslide inventory, subsurface property	Atterberg limit, clay mineral	infinite slope stability analysis	(1983)
Landslide inventory, subsurface property	grain size, root effect	infinite slope stability analysis	(1982)
Observation of debris flow	physical property of slurry		(1981)
Subsurface property	geotechnical tests	infinite slope stability probability analysis	(1982)
Soil survey, soil water status			

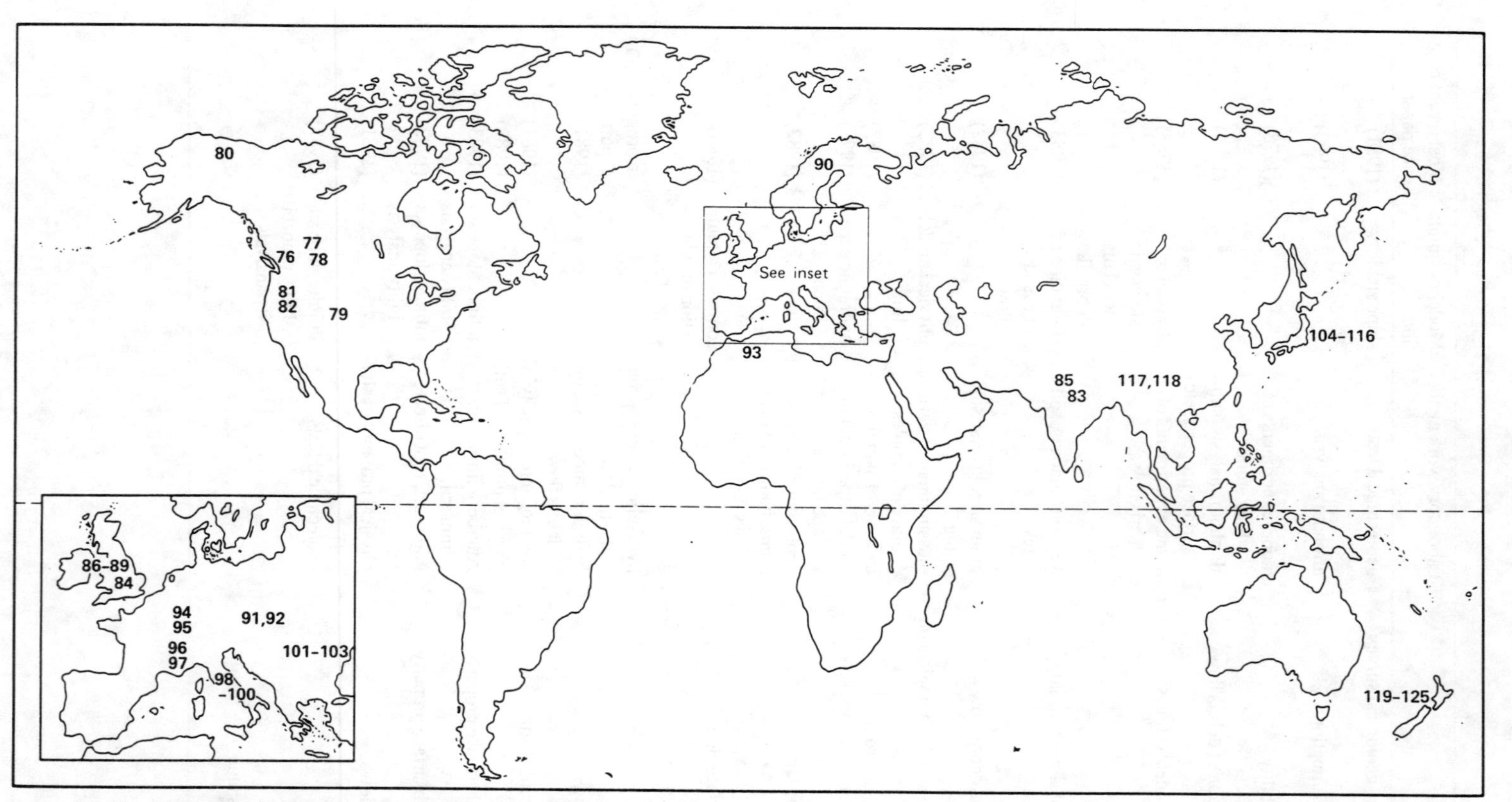

Figure 3.1. Location of rapid mass movement experiments reported.

3.1.2 *Causes of mass movement*

Before introducing a field experimental approach for rapid mass movement, it is necessary to review the general information about its occurrence and characteristic motion.

All slope materials have a tendency to move down slope under the influence of gravity, but this tendency is counteracted by shearing resistance. Only when the downward force exceeds the resistance, some types of failures occur and materials will be displaced with various speeds to a new equilibrium position. There are two kinds of causes of mass movement which make the downward force to exceed the resistance, i.e. external and internal ones. The external causes produce an increase in the downward force or in shear stress within the materials, while the internal causes produce a decrease in resistance to prevent motion or a decrease in shear strength.

Typical external causes are briefly described as the following:

a) Geometrical change in slope shape, such as an increase in gradient of slope by orogenic motion, stream incision or artificial excavation in slope foot.

b) Loading at upper slope or unloading at lower slope by natural or artificial actions.

c) Shocks and vibration, for example, occurrence of earthquake, explosion or machinery vibration.

d) Changes in hydrological conditions, for example, heavy rainfall, rapid snow melting or rapid drawdown of ground water.

Typical internal causes are briefly described as the following:

a) Weathering and progressive failure in subsurface slope materials over a long period.

b) Seepage erosion or piping under special subsurface hydrological conditions.

Rapid mass movement involves two distinct physical events, i.e. slope failure and rapid movement of failed materials. The static or dynamic study on the initiation of slope failure is well developed in the field of soil (or rock) mechanics and introduction of their results is neglected in this report. But rapid movement of slope materials over a long distance after their collapse has yet to be explained comprehensively from a physical view point. The special phenomenon of fluidization or air-layer lubrication can play an important role for reduced friction in some types of movement and Bagnold's theory on cohesionless grain flow (Bagnold 1957) can give a reasonable explanation for some kinds of mass motion of grains with high concentration in water or gas because much reduced internal friction is caused by elastic collisions among grains.

3.1.3 *Classification of rapid mass movement*

In real slopes, rapid mass movement occurs under complex combinations of external and internal causes and it takes various patterns of failures and motions of slope materials. Then, some classification is needed for practical treatment of mass movement to investigate the phenomenon more concretely and to take more effective counter-measure for some hazards caused from various motion patterns. For the present, we will adopt the simplest classification which defines the most commonly appearing types of mass movement under the five headings of fall, topple, slide, flow and creep. The first four belong to rapid mass movement in most cases and the last one, creep, is usually treated as a slow mass movement, even though some accelerative creeps may bring about slope failures and they can change into rapid slide or flow suddenly.

Still more, slide is often classified into types, rotational and parallel ones and the rotational slide is called 'slump' usually. Then, five terms, falls, topples, slumps, slides and flows are adopted in Table 3.1. For reasonable planning of field experiments, we have to select most appropriate method taking the characteristic motion of each type into consideration.

In order to obtain the best results from mass movement studies, a synthetic approach including all effective methods is needed, and besides field work, some items about laboratory work, modelling and simulation are listed in Table 3.1.

In the following sections, various methods for study of rapid mass movement and results obtained from the methods are introduced.

3.1.4 *Various approaches for study of rapid mass movement*

In field experiments on rapid mass movement, field observations are needed to observe, measure and record various processes as quantitatively as possible. Observation methods should be selected according to the purpose of experiments which are often specific to the type of rapid mass movement, and each geomorphologist should establish his own observation method according to his purpose and research circumstance. But there are some common objectives of field observation for which many geomorphologists are working in various places in the world to solve somewhat common problems about rapid mass movement. The following is a list of common objectives of field observation for studies on rapid mass movement which are selected from the contents of Table 3.1 and my own experience, and some explanations are added to describe the reason for selection.

Slope morphometry

Topographic changes in slopes, valleys and fans are important objects of field observation for the study of rapid mass movement, because they are the direct results of rapid mass movement and the best clues to find the aggregate effect of various geomorphological processes. In every field experiment, qualitative or quantitative observation of topographic change is always necessary, though the scale of observation in space and time is different depending upon the purpose of each experiment.

In situ measurement for slope failure study

Mass movement starts from a slope failure which is controlled considerably by the strength of slope materials and the moving pattern of the materials after failure depends upon the rheological behaviour of the materials. Then, physical properties of slope surface should be investigated. Reasonable choice of test points is very important to investigate physical character of whole slope or local special spots, because physical properties of slope surface are usually not uniform along a slope surface or orthogonal to the surface.

Hydrologic factors

Most rapid mass movement has a direct relation with special hydrologic factors, for example, heavy rainfall or rapid snow melting. The distribution and movement of groundwater near the slope surface also control the mass movement phenomenon strongly. Then continuous observation on hydrological conditions at specified slopes is necessary to predict the initiation of mass movement and to investigate the movement of materials. Special care has to be taken to observe rapid change in hydrological conditions and some original methods are required for this special purpose different from general problems on water balance study.

Dynamic phenomena

Some efforts to observe or measure the movement of slope materials have been made to clarify the physical role of individual mass movement processes (fall, topple, slide, slump and flow) in producing topographic changes. But our object of observation is the movement of materials in collective or massive form and not the motion of individual solid particles which is outside of this report. Most investigations are carried out simultaneously measuring the physical properties of slope surface and exogenetic actions including natural and artificial impacts to study quantitative relationships between exogenetic actions and mass movement.

Besides the above mentioned field work, laboratory work is needed to prosecute synthetic study of rapid mass movement.

Laboratory work, modelling and simulation

Laboratory work is carried out in many studies of rapid mass wasting as shown in Table 3.1 and is roughly classified into four kinds – mapping, geotechnical test, modelling and simulation. Recently very useful equipment including high speed computers are being utilized for laboratory study of geoscientific and geotechnical problems and the most effective method should be selected carefully taking the special character of each method into consideration. Typical examples getting successful results in laboratory study are introduced.

Hillslope evolution

Rapid mass movement changes landscape most actively at many places in the world and plays an important role in hillslope evolution. Short term and long term prediction of change in hillslopes can supply very useful information for geoscience and environmental science. A synthetic approach is needed to study hillslope evolution quantitatively through field and laboratory works, physical and theoretical modelling, and numerical simulation.

3.2 SLOPE MORPHOMETRY

General techniques to survey topography and to make a map are introduced in text books of surveying and cartography. But special care has to be taken in studying rapid mass movement to detect small and rapid changes in topography exactly in short time. Some special methods to observe topographic changes for the above purpose are introduced in the following.

3.2.1 *Slope form*

Exact and repeated surveys of slope form are most important for the study of rapid mass movement, and surveying or mapping from airphotos are standard procedures. But a

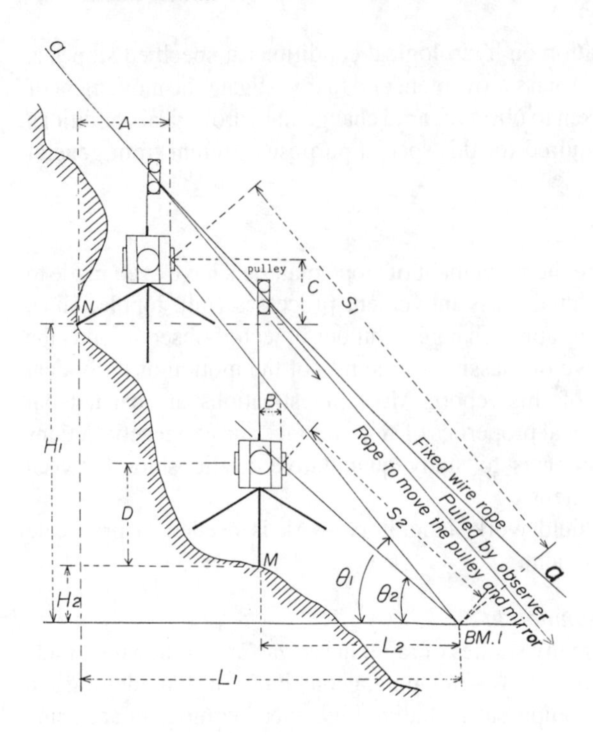

Figure 3.2. Survey technique for slope profiling with a new reflecting target which can be moved by a surveyor pulling the rope; see text for detailed explanation.

simple and inexpensive method is needed for special surveys which can be practised easily by geomorphologists.

One such useful field technique for profiling precipitous slopes is introduced by Churchill (1979). In his method, the positions of target points are determined with vertical angle θ and straight line distance S from an instrumental station BM. on the slope foot with a transit or theodolite to measure θ and with an optical range finder to measure S as shown in Figure 3.2. Profile of the slope connecting target points can be determined from simple trigonometry. This method is especially useful where surveyors cannot access target points. Measurement error is dependent upon instrument accuracy, particularly the accuracy of the ranging device when an optical range finder is used.

To reduce such measurement error, Yokoyama (1983) devised a new technique using an electromagnetic wave range finder combined with a theodolite and special reflecting mirrors, and he obtained good results for profiling steep slopes. The principle of surveying is the same as that by Churchill, but as a target, a special reflector can be set in any point along the slope by a surveyor who controls the target position with a wire rope and pulley system from the slope foot as shown in Figure 3.2. To get a reflecting electromagnetic wave from the target to the range finder at the slope foot regardless of the position and inclination of the target, a special reflecting target was devised which consists of eight reflecting mirrors facing various directions and has eight feet for touching slope surface. This target was designed to make a survey of profile possible even along a slightly overhanging slope. The instrumental error for the distance between the target and the range finder of several hundreds of meter length can be confined to ± 1 cm. An example of slope profiling by this method at a test field is shown in Figure 3.3 with small circle marks.

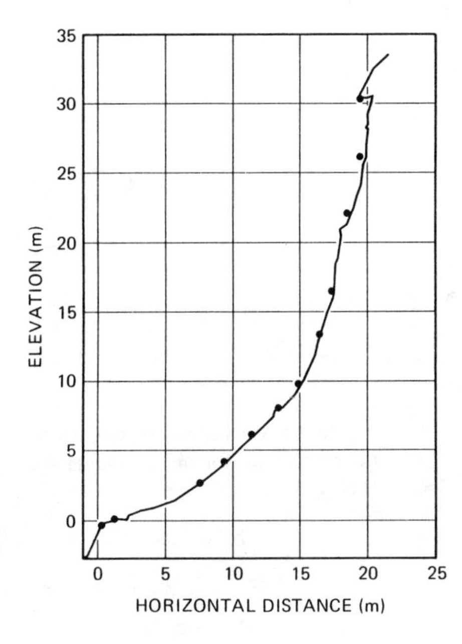

Figure 3.3. Comparison between slope profiles obtained by different methods; circles: by a range finder and theodolite using a special target, line: from a pair of terrestrial photographs.

To check this result by comparing it with the other data from an independent method, a topographic map of the same slope was made from a pair of stereo-photos which were taken with a camera Wild P-32 on a base line of 14 meter length parallel to the slope foot and one hundred meters from the slope. The topographic map with scale 1/100 was made with stereo plotter Wild A-7 from the pairs of photos and the measurement error seems to be less than 10 cm. The profile depicted from the topographic map by stereo-photographing along the same cross section is shown as the solid line in Figure 3.3. Results from these different methods almost coincide with each other except in very steep regions where a small shift of the surveyed line between two methods may bring about a large height difference.

Rapid retreat of steep slopes is a phenomenon of great interest in geomorphology, and this new technique using a reflecting target seems to be useful for field experiments on rapid mass movement.

On the other hand, mapping from stereo photos also is a useful method where the topography is changing so fast that other surveying methods cannot record the topographic change rapidly enough.

3.2.2 *Micro-topography*

For field experimental work on rapid mass movement, some special techniques are often needed to detect a micro-topographic change exactly with a simple and inexpensive instrument which can be operated by a geomorphologist. One such technique was devised by Ishii (1981) to survey a micro-topographic change by one surveyor using a new instrument. His instrument consists of two legs *AC*, *AD* of equal length, one bar *BE* and a string *AG* hanging a weight *G* as shown in Figure 3.4.

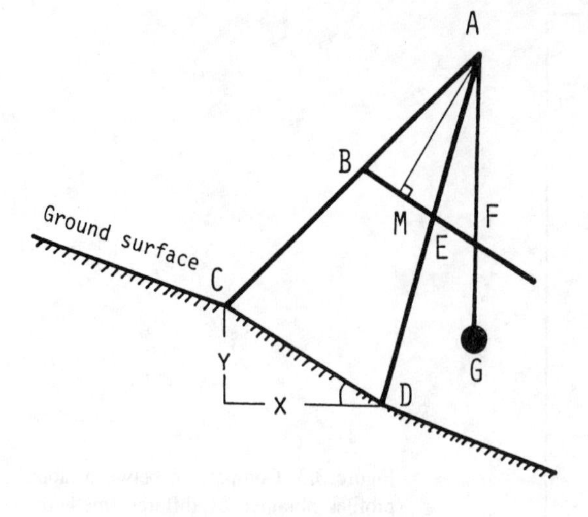

Figure 3.4. Conceptual representation of instrument for microtopographic survey.

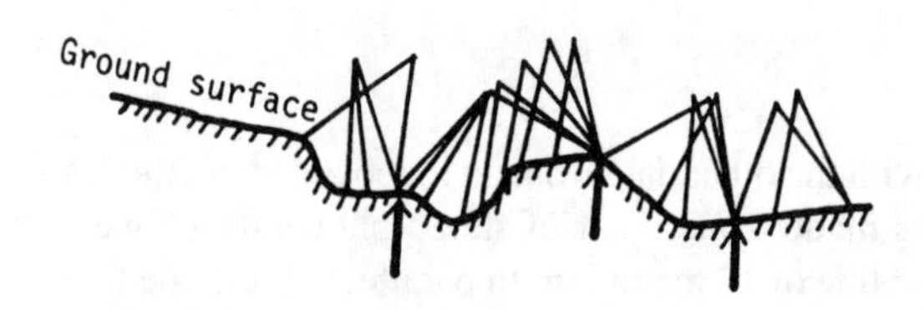

Figure 3.5. Measuring method with the instrument in Figure 3.4. Arrows show the points fixed temporarily.

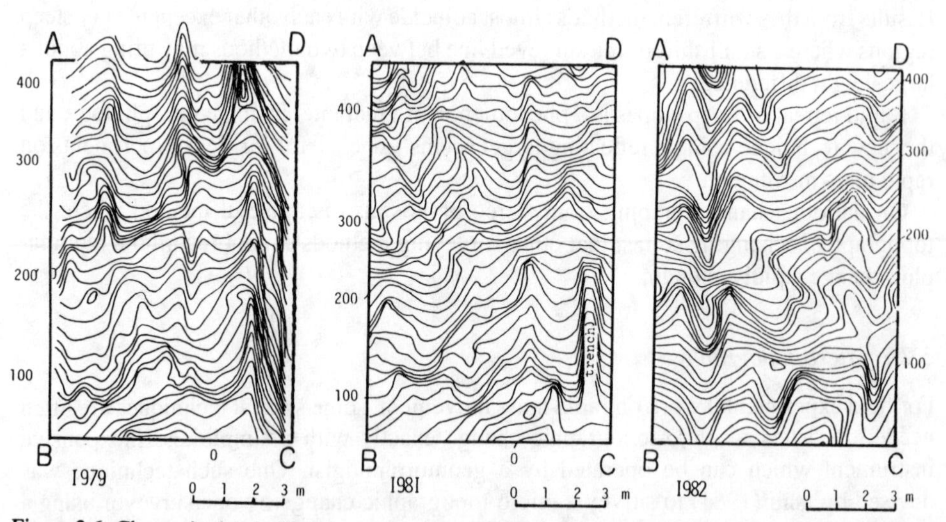

Figure 3.6. Change in debris flow deposits at the test field. Individual quadrangles indicate the same place which is the usual depositional area of debris flows and also had been the foot of talus before 1970.

If we have a constant length of *AB* and *AC* in construction of the instrument and read the lengths *BE* and *BF* with a scale on the bar in surveying, we can determine *X* and *Y* using a simple calculator.

The two legs are set on the ground surface as shown in Figure 3.5, fixing one leg at one point and moving the other leg to any measuring points along ground surface at an appropriate interval.

As an example of results obtained by this method, microtopographic changes by debris flow deposits which were observed by Ishii at Ashio test field, Kanto District in Japan are shown in Figure 3.6. The micro-topography of this field was surveyed repeatedly in 1979, 1981 and 1982 to investigate the depositional pattern by small debris flows which had occurred in 1975, 1977, 1981 and 1982.

3.2.3 *Changes in ground surface level*

When the movement of slope materials is limited within a thin ground surface layer, a simple observation of change in ground surface level can give useful information about the mass movement phenomenon. A rod or stick driven into an immovable layer is usually used to measure the change in ground surface level and to detect the rate of erosion or deposition by mass movement. But normal measurement with a rod or stick shows only the net change in ground surface level though most of the actual changing process goes on with a combination of scouring and deposition during one mass movement period, for example, initial scouring and succeeding deposition or vice versa. New techniques are needed to measure not only the final ground surface level but also the highest and lowest level during an observation period.

One such technique was devised by Leopold (1967) and a similar combination of rod and ring was tested to investigate a depositional pattern of debris flows at a fan by Okuda et al. (1979). The steel ring can settle down along the rod to the bottom of the disturbed layer during a scouring stage, then deposits cover the ring during the successive depositional stage. We can find the top level of undisturbed layer from the position of the ring by digging it out and find the final level of deposition from the last ground surface level. By this method, it was found that deposition of debris flows sometimes occurred after initial scouring of several tens of centimeters depth in some parts of the lower fan.

In the reverse case of initial deposition and subsequent scouring, the highest level of ground surface cannot be detected by this method. But the maximum water level gauge, which records the maximum water level by colour change of a special material getting wet with water, can be used to detect the highest level during a mass movement event because the moving materials usually contain some water.

3.3 IN SITU MEASUREMENT FOR SLOPE FAILURE STUDY

For field experimental study on rapid mass movement, various measurements of physical properties of slope materials are needed to investigate the physical state of a slope before and after mass movement. The physical properties in this section are used in a wide sense including all mechanical, physical and geochemical properties which have some relation with rapid mass movement, but in the following explanation measurements are limited to special properties in relation to the strength of slope materials. Measurement of slip

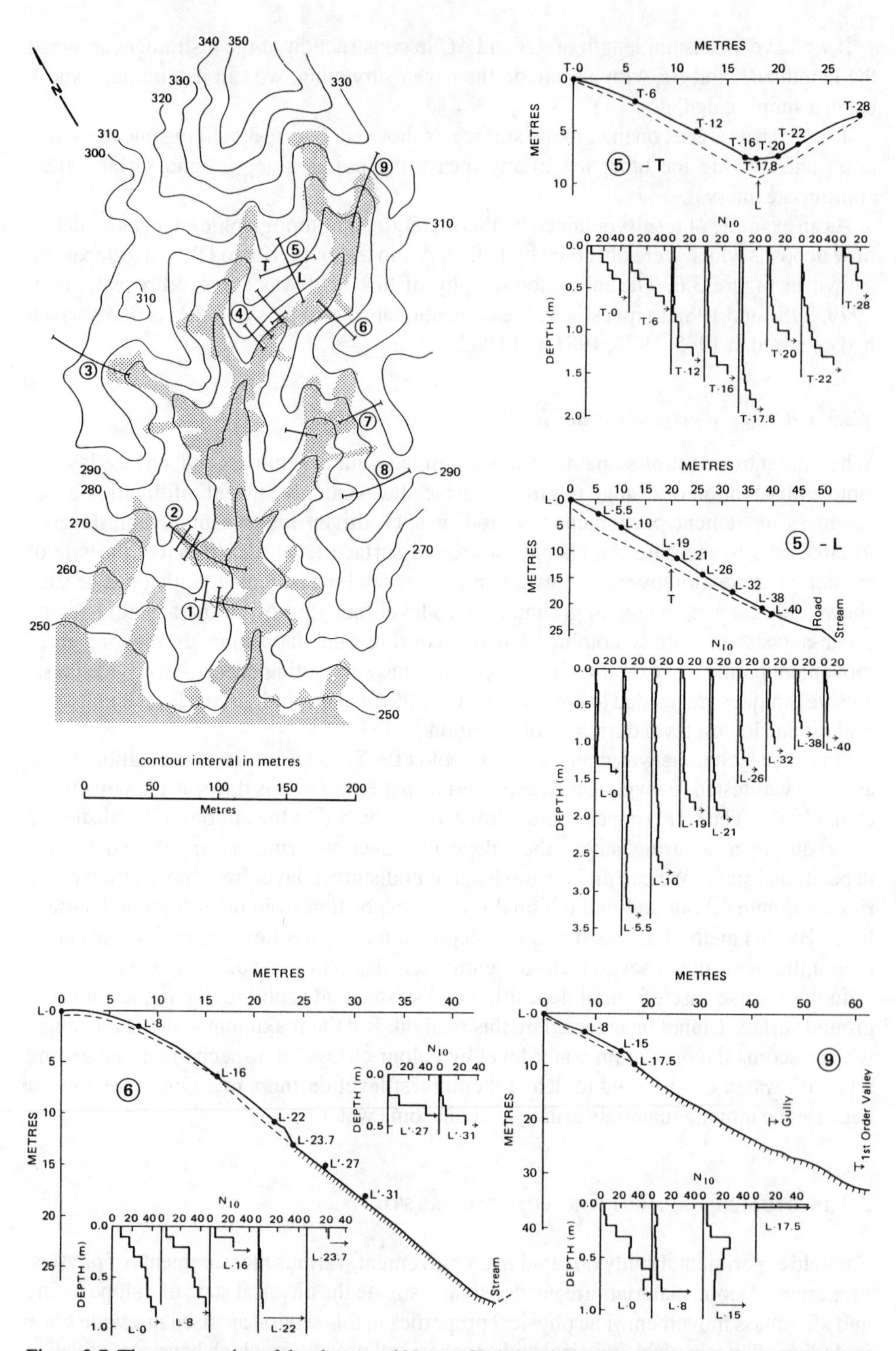

Figure 3.7. The cross section of the slopes and N_{10} profile in weathered granite hillslopes in the left map at Obara Village, Central Japan where many landslides were caused by a heavy rainfall in 1972. The arrows in N_{10} profiles represent the depth where N_{10} exceeds 50.

surface is also an important work for slope failure study which should be prosecuted only in the field and depends on the distribution of physical properties of slope materials.

Only in situ measurements, which are simple and can be used by geomorphologists for field experimental purposes, are introduced in this section and general explanation on various methods should be referred to papers in soil mechanics (e.g. American Society of Civil Engineers, 1975).

3.3.1 *Subsurface sounding*

'Sounding' means an investigation of mechanical properties of subsurface materials to insert a rod with some resistant head (cone, vane or sampler), and to measure resistance with penetration, rotation or pulling up of the rod. If the resistance changes along the vertical line to the ground surface, the relative difference of mechanical properties at various depths is an important factor in relation to potential mass movement.

Soundings are classified as static or dynamic according to the method used to insert the resistant head into the ground. In a static type, the resistant head is pushed into the ground by static loading with weights (of iron block or human body weight). This type is used more effectively in soft ground. In a dynamic type, the resistant head is driven into the ground by impact force with dropping a specified weight from a specified height on the upper end of the rod. This type is used more effectively in hard ground.

For field experimental study on rapid mass movement, the dynamic type is more convenient in most slopes except in very soft deposit materials of high water content. In the former type, the strength of ground materials is expressed by the static force to push the head into the ground under a constant condition, while in the latter type, the strength of ground materials is usually expressed by the count number of dropping of weight with a constant energy to penetrate the head through every specified depth and the count number is called N-value. For example, in Japan, N-value in standard test is defined as the number of impacts by dropping the weight of 63.5 kg from the height of 75 cm required to penetrate 30 cm into the ground. In the standard test, the head part serves both as a resistant head and a sampler which can sample ground materials by pulling out the head from the ground after inserting test. The head has a standard specified size.

Sometimes, a kind of N-value, N_{10} is used. N_{10} is defined as the number of impacts required to penetrate successive 10 cm depths to investigate more finely the vertical structure of the shallow near-surface layer.

From N-values measured by a sounding, a bulk density or friction angle of ground materials can be estimated directly by some empirical relations derived from laboratory tests. From adequate sounding tests, a vertical distribution of relative strength of ground materials and the overlapping of different materials are recognized by the vertical distribution of N-value.

In the sounding of ground containing large stones, special care is needed to judge the result of a sounding taking a local extraordinary high value into consideration.

Some examples of sounding for field experimental study on shallow landslips in Japan are described in the following. By a sounding at slopes of Mt. Rokko, northern part of Kobe City, where granitic rock has been strongly weathered and shallow landslips have been caused frequently with heavy rainfalls, a positive correlation between N_{10}-values measured with a penetrometer of Doken type (dropping a weight of 5 kg from 50 cm height) and dry bulk density of slope materials was found. A superficial layer N_{10}-value < 10 is most apt to fail and to slide in the midst of heavy rainfall (Tanaka and Okimura 1976).

On another slope of weathered granite in Obara Village, western part of Aichi Prefecture, Central Japan, where a heavy rainfall caused many landslides in 1972, soundings with a penetrometer of Doken type were carried out to examine a relationship between the landslips and the subsurface structure of the slope by Iida and Okunishi (1979). A part of the results is shown in Figure 3.7 illustrating the vertical distribution of N_{10}-values. Broken and chained lines in this figure show the depths where N_{10} exceeds 10 and 50, respectively. As illustrated in the figure, at the middle and lower segments of the slope where landslides occurred most frequently, the soft layer with $N_{10} < 10$ is very thick, while the transient zone between soft and hard layers with $10 < N_{10} < 50$ is extremely thin. Then the subsurface structure can be regarded approximately as two-layered pattern which means that a hard bed rock underlies a strongly weathered and soft layer with a sharp boundary between them, and this boundary seemed to play the role of slip surface in heavy rainfall.

As explained by these examples, a sounding is very effective to survey the subsurface structure in slopes and to estimate the depth of slip surface from relative strength of slope materials in shallow landslips. But in usual sounding with simple equipment, sounding depth is limited within several meters below the ground surface and for study on deeper slope failure, some other methods to investigate the deeper subsurface structure are needed as explained in the following.

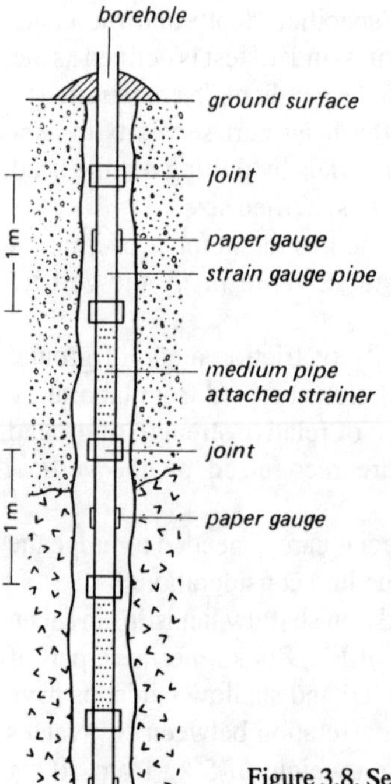

Figure 3.8. Setting of strain gauges on a pipe (from Japan Society of Landslide 1980).

3.3.2 *Locating of deep slip surface in landslides*

For study on a large scale landslide, locating of deep slip surface and determination of its shape is important to investigate the mechanism of sliding process and to predict the initiation of sliding. Though most landslides with a deep slip surface belong to a slow mass movement, in some large landslides especially those caused directly by a heavy rainfall or a big earthquake, rapid movements occur and some measurements of this process should be treated in this section.

Hutchinson (1982) reviewed the main methods of discovering the shape and locating the depth of slip surface in landslides after stressing the importance of surveying them, regardless of the depth to slip surface. From practical engineering view points for landslide prevention also, various kinds of methods of locating slip surfaces have been developed to check slope stability quantitatively. Then, in this section, there is a brief description of some methods to investigate slip surfaces especially emphasising usefulness for field experimental studies of rapid mass movement. General information about the investigation of slip surfaces should be referred to engineering text books (e.g. Zaruba and Mencl 1982).

When a sliding mass is moving actively the cross-sectional shape of a borehole penetrating an active slip surface will be transformed by shearing action in a short period. The depth of the slip surface can then be detected by the location of the changing cross section. Sometimes, the position of the changing cross section can be observed directly by throwing a light beam into the borehole and the depth can be measured by a tape.

After a slight shear, a hook attached to the lowest end of a measuring tape is effectively used to measure the depth of a somewhat overhanging step in the borehole by hooking. After continuous shear and complete blocking of the cross section, the depth of slip surface can be measured only by hanging a tape.

When a sliding mass is almost stationary, a shear rate of cross section of borehole is very small or zero even in the slip surface, the above mentioned method cannot be applied and other methods are needed to locate the slip surface. In Japan and some other countries, as a simple and inexpensive method, a set of strain gauges attached to a flexible pipe, is used widely to detect the depth of slip surface and to record the shear continuously for a long period. The measuring principle is explained by Figure 3.8. One set of pairs of strain gauges is mounted on outside surface of a flexible pipe (in most cases of vinyl chloride) facing each other at the respective depth of measurement and wired to an ohmmeter. After inserting the pipe into a borehole, the pipe will be bent by the shear effect near the slip surface and the one gauge will be extended and the other one in the opposite side will be contracted at the bending part of the pipe. Then the change in electric resistance of gauges doubled by the combination of coupled gauges can be detected by an ohmmeter at ground surface. By successive reading of the resistance of each pair of gauges the position of the bending part of the pipe can be located and by repeating such measurement at a specified time interval, the shear along the slip surface can be followed for a long period. An example of a record of extraordinary change in strain calculated from electric resistance which was observed in a landslide area in Japan is shown in Figure 3.9. A sharp slip surface and rapidly proceeding shear are found in this figure. From such a continuous record, a relationship between rainfall (or snow melting) and accelerating or decelerating rate of slip motion can be investigated which is useful for a prediction of slope failure or initiation of rapid motion.

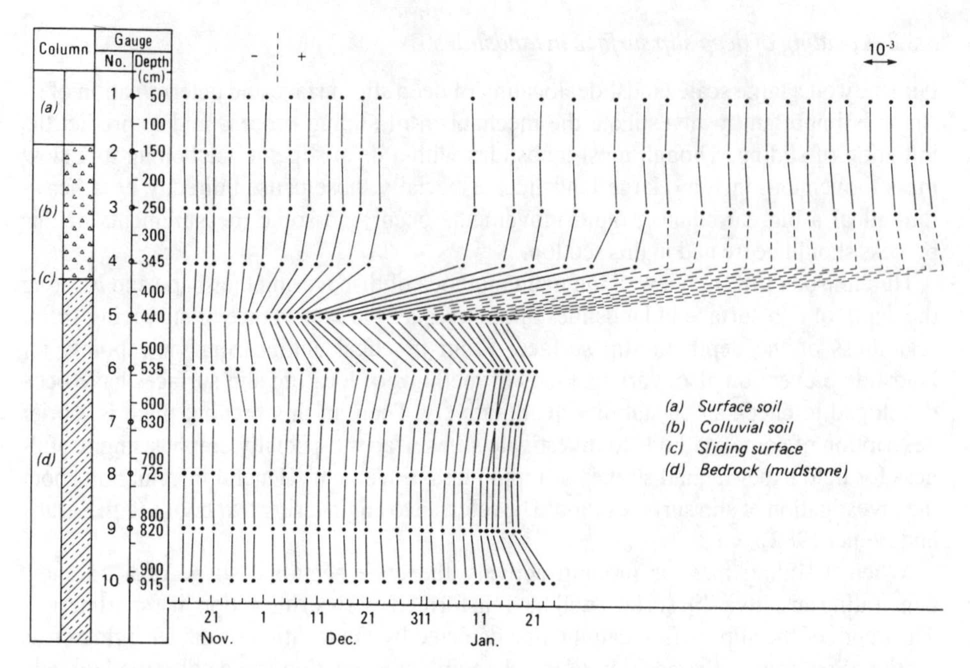

Figure 3.9. An example of measurement result of strain gauges observed at a landslide area (from Japan Society of Landslide 1980).

In a completely stationary slide, change in a borehole cannot be detected even by the above method. The depth of a potential slip surface where a slip had occurred sometime in the past and will occur in the future has to be located by another method. For this purpose, some methods of geophysical exploration are utilized effectively to investigate vertical distribution of physical or chemical properties of underground materials and to locate a sharp transient zone of the properties which corresponds to the potential slip surface. For a full explanation the reader should be referred to a text book of applied geophysics (e.g. Milton 1976).

3.4 HYDROLOGIC FACTORS

Observations are needed to investigate a relationship between specified hydrologic factors and failure of slope materials and to predict the occurrence of rapid mass motion quantitatively.

One of the most important hydrologic factors is rainfall as direct input which brings about slope failures in many cases. And the other one is ground water level, a measure of hydrological response of slope to input water, and it is a more direct factor than rainfall to control hydraulic conditions for slope stability.

As an intermediate process between rainfall (snowmelt) and groundwater supply, infiltration plays an important role for slope failure and rapid mass wasting. These three hydrologic factors – rainfall, infiltration and ground water level – were selected in this section as direct objectives of observation for field experimental study on rapid mass movement.

3.4.1 *Observation of rainfall on slope surface*

In general, rain water flux does not arrive at a slope surface perpendicularly to it and rainfall intensity (the water flux per unit time per unit area along the slope surface – R_n) is dependent on the falling direction of rain drops, the direction normal to the slope surface and the absolute flux of rain water (R – water flux per unit time per unit area perpendicular to the direction of rain drop motion).

From three dimensional geometry,

$$R_n = R\,\{\sin\phi \sin\phi' \cos(\theta - \theta') + \cos\phi \cos\phi'\}$$

where the direction of rain drop motion is given by azimuthal angle ϕ and zenith angle ϕ' and the direction normal to the slope surface by θ and θ' respectively. If the rainfall flux is captured by a normal opening of raingauge which is set horizontally, the flux measured by the raingauge R_0 is given by the next equation

$$R_0 = R\cos\phi$$

Therefore, there is some difference between R_n (water flux on a slope surface) and R_0 (water flux on a horizontal opening) and some care has to be taken to estimate the water flux on a slope surface from a record of a normal raingauge. Then the directions of rainfall flux (the direction of rain drop motion near the slope surface) and of the normal to the slope surface have to be measured simultaneously to estimate the water flux which would be supplied to the slope surface. The direction of the normal to the slope surface can be easily determined by a transit survey, but the direction of rainfall flux cannot be determined by a normal raingauge.

In our field experiment at Mt. Yakedake, a special vector rain gauge developed by Sakanoue (1969) and improved by our group was used to estimate the rainfall flux to the specified slope surface. The vector rain gauge has five receiving planes which are facing mutually perpendicular directions, i.e. vertical upward, northern, eastern, southern and western directions to catch the rainfall flux from the five directions and to collect the rain waters by five tipping buckets for each direction separately. Then the direction of rainfall vector (θ, ϕ) and absolute value of water flux (R) can be determined from the composition of three components of rainfall, one vertical and two horizontal components at every observation period.

With this special equipment, the characteristic pattern of rainfall vector was observed near the mountain top where a strong wind with large horizontal component of speed often blows, and rainfall direction deflects largely from vertical direction. The difference of rainfall flux between different slopes cannot be neglected even if the distance between the slopes is small in mountainous regions under special conditions with strong wind and heavy rainfall, and such a difference of hydrological situation may bring about the difference of slope failure incidence.

3.4.2 *Observation of infiltration*

In most rapid mass movement, infiltration of water into ground from rainfall or snow melting has an intimate relation to initiation of movement and features of the motion, so measurement of infiltration rate or infiltration capacity is one of the most important indices of physical properties of slope materials. 'Infiltration rate' is a measure of water quantity infiltrating per unit time (usual unit mm/hr) and 'infiltration capacity' is the largest possible infiltration rate at which ground materials can absorb rain water as it falls.

Infiltration is a complex physical process which is controlled by mutual interaction among solid particles (fragments), water and air through very small pores. Various attempts have been made to measure the infiltration rate correctly through solving some technical problems. For in-situ measurement of infiltration capacity, an infiltrometer is used which supplies water on the ground surface under controlled conditions and infiltrates the water into the ground. It is usually classified into three types, i.e. flood-type, flow-type and rainfall-simulator-type.

The flood-type of infiltrometer is the most popular and it supplies water from a tank to the ground surface keeping a constant head over a small area inside a metal circular cylinder. A convenient infiltrometer of this type was devised by Hills (1970).

The flow-type of infiltrometer supplies water from the upper side of a test area and receives the water at the lower side of the area and the discharge rates of water of the upper inflow and the lower outflow are measured by a suitable method. The infiltration rate is calculated from the difference of the two discharge rates divided by the slope area which is covered with the sheet-flowing water. This type of infiltrometer is especially suitable for measuring infiltration rate at a slope where a constant water depth over a test area cannot be held by horizontal water surface in a cylindrical frame of flood-type infiltrometer. A convenient portable flow-type of infiltrometer was developed in Japan by Sato et al. (1950).

A rainfall simulator type of infiltrometer differs from the flow-type in that it supplies water by a spray of simulated rainfall, not by surface sheet flow. Therefore, it can measure the infiltration rate under more natural conditions of rain drops beating the ground surface and surface flow is very thin. Various spraying methods are adopted by various hydrologists, but it is difficult to produce rain drops of size and of fall velocity similar to natural rain drops. This type of infiltrometer is a high cost and complex operation and is not suitable for field work.

Infiltration capacity usually changes from place to place owing to the variety of local properties of ground materials and it is necessary for an extended area survey to measure the capacity at a number of testing points in order to obtain a representative value of infiltration capacity over the area. Especially for study of rapid mass movement, by contrast with general water balance problems, some special care has to be taken to find special concentrated sources of groundwater not only by using infiltrometers but also by, for example, observation of distribution of groundwater head or tracing of groundwater flow.

3.4.3 *Observation of ground water level*

Observation of ground water level on slopes for field experimental study on rapid mass movement needs special care because the change in water level occurs suddenly and rapidly under extraordinary hydrological conditions. The water level gauge has to be devised to measure a large range of level change with a rapid response, to continue the measurement under conditions of slight deformation of boreholes in which the gauge is set, and to record the level by telemetering over long distances. Still more the sensor of level change must be cheap because many gauges are usually distributed in an experimental field to investigate the change in the ground water level simultaneously over an area at many observation points and sometimes there is a risk of losing the sensors by filling up of boreholes by mass movement.

Okuda et al. (1976) developed a new type of groundwater level gauge to measure rapid change in the water level on a slope surface or a valley bed during a heavy rainfall and

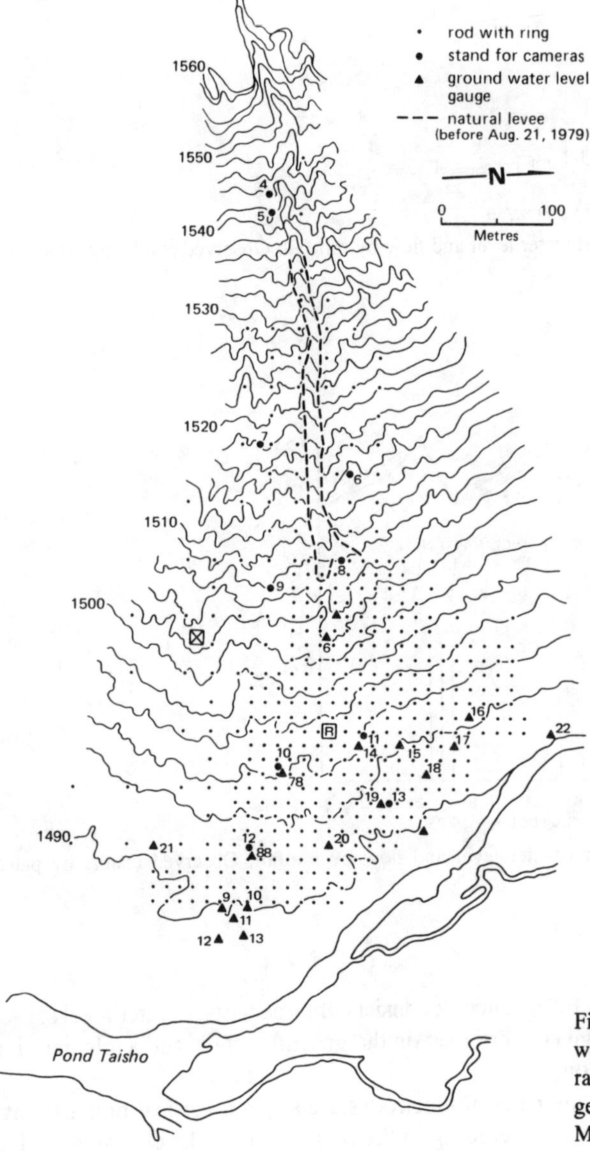

Figure 3.10. (a) Changes in ground water level and flow by rainfall: Arrangement of ground water level gauges at Kamikamihori fan at the foot of Mt. Yakedake.

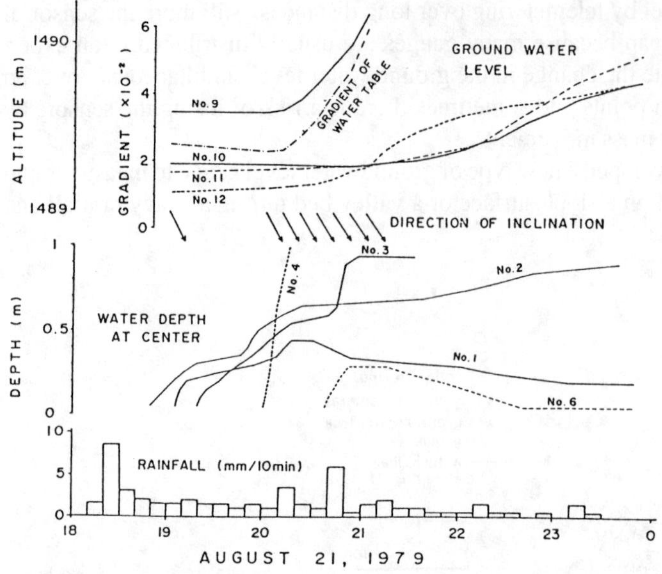

Figure 3.10. (b) Changes in ground water level and flow by rainfall: Observed results by continuous recording.

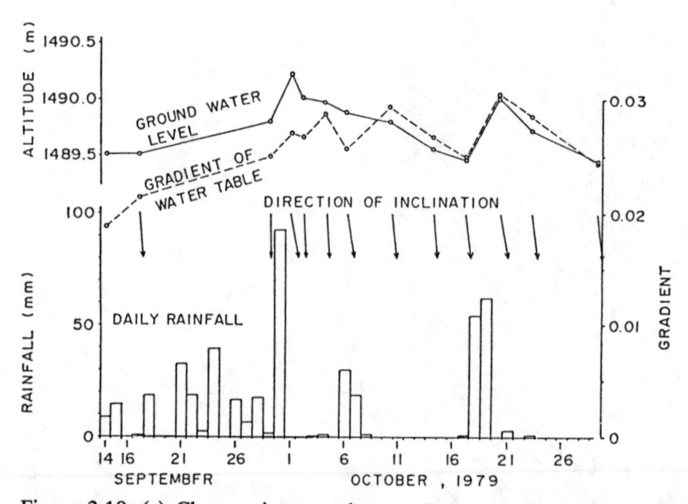

Figure 3.10. (c) Changes in ground water level and flow by rainfall: Observed results by patrol measurements.

even in the midst of debris flow occurrence. It consists of three parts, a water level sensor set in a borehole, a level-voltage converter set on the ground surface and a telemetering recorder in an observation station.

In the water level sensor, many pairs of electrodes are set in a vertical pipe of vinyl chloride at suitable height intervals according to the range of water level change and to the accuracy needed for the study. An electric circuit including a pair of electrodes is on or off depending on the position of the electrodes in or out of ground water column and a change in ground water level is converted to a change in output voltage through the change of resistors in the voltage converter. Continuous recording is possible by measuring the output voltage and long distance telemetering (for several kilometers) is possible.

A part of the records of ground water levels observed by Okuda et al. (1980a) at the fan of Kamikamihori is shown in Figure 3.10. The gauges were distributed in the fan as illustrated in the Figure 3.10a to investigate the change in ground water levels and the direction of ground water flow in the midst of heavy rainfall at the lower fan. A group of gauges No. 9, 10, 11, 12 and 13 was set to measure the gradient of ground water level i.e. the direction of the ground water flow at the specified region. A rapid rise of water levels was found in the Figures 3.10b, c from the continuous record.

Another example of ground water level survey which was useful for prediction of slope failure was introduced by Okimura (1983) in the field experiment on landslips caused by fissure water in Mt. Rokko near Kobe City.

3.5 DYNAMIC PHENOMENA

Observation of motion of slope materials in the midst of rapid mass movement is more difficult than that of topographic change between, before and after some geomorphological events, because the place and time for field observation cannot be determined prior to the occurrence of mass movement, and instruments necessary for observation or recording of the motion of materials are generally more complex and expensive than those for normal topographic surveying. Special care must be taken to select the place, time and instruments according to the expected phenomenon and some examples for the observation of motion of different phenomena are introduced.

3.5.1 *Observation of 'fall' phenomenon*

Fall is a type of rapid mass movement which occurs often on steep slopes and the terms 'rock fall, soil fall and debris fall' are usually used depending upon the kinds of the falling materials. The 'fall' occurs suddenly in most cases without any precursor, and direct observation of its moving state is very difficult. But careful and long-term observation of special slopes where 'fall' phenomenon occurs frequently, can catch the actual situation and give us useful information about physical conditions controlling this phenomenon.

Rock fall has been investigated by some geomorphologists (Rapp 1960, Luckman 1976) taking climatic and geologic factors into consideration, and Gardner (1983) has carried out field observation on rock falls over a seven-year period and demonstrated their frequency and distribution at the Highwood Pass area, Canadian Rocky Mountains. In those studies, most rock falls were identified and located by the sound they produced. A new observation method for rock fall was developed by Suwa and Okuda (1983) using an automatic camera at a fixed and small area in a valley in the Japanese Alps. They carried out a field experiment on rock fall to record the displacement of individual rock fragments continuously for two years at the sidewall in Kamikamihori Valley on the eastern slope of Mt. Yakedake, Central Japan. A constant interval shot camera was set on

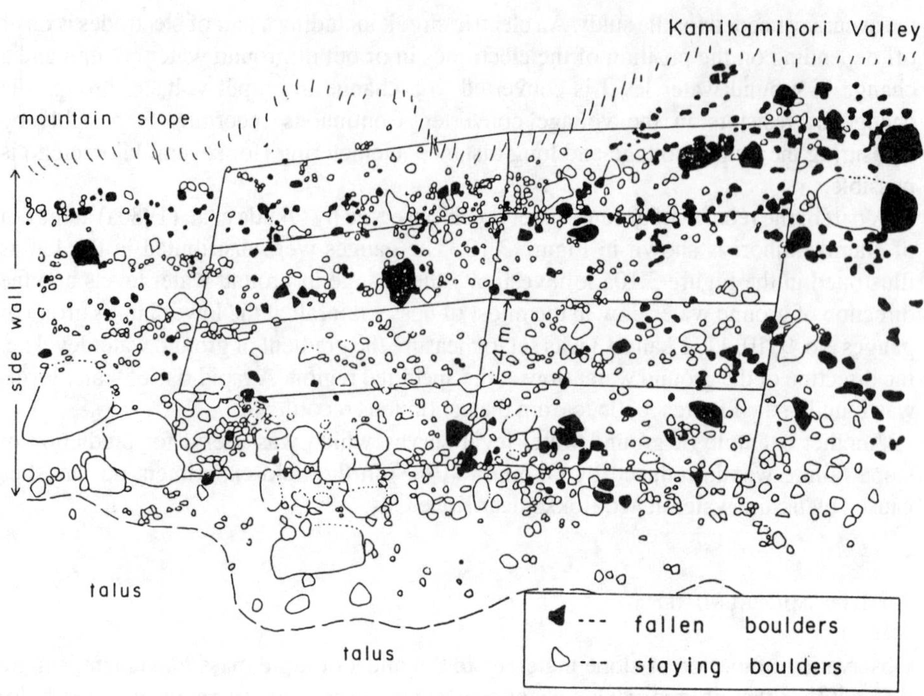

Figure 3.11. Rock fall from a flat sidewall which occurred in the observation period June 17, 1981-Sept. 29, 1982.

the top of the opposite sidewall to catch the object, a specified area of valley wall and bed in the fixed visual field of the camera, and successive photographs were taken automatically every two hours in daytime to record the change in surface pattern of the valley wall caused by rock fall and the change at the valley bed by accumulation of fallen rock fragments. With a special film magazine, film exchange only once in a month was sufficient to continue the automatic recording. Several targets of 20 cm × 20 cm square plates were fixed to the ropes hanging over the valley wall to put the standard scale of length in the visual field of the camera. By comparison of successive two photo prints enlarged in the same size, the fall and settling of rock fragments larger than 10cm in diameter can easily be recognized, and a part of the results of this field work is illustrated in Figure 3.11.

Besides the continuous photographic recording, the meteorological data on air temperature and rainfall have been collected to study the controlling factors for rock fall, and topographic surveyings have been repeated several times to investigate topographic change along the valley including special areas where the above mentioned photographic recordings were performed.

Some field experiments on 'fall' phenomenon were carried out to observe directly the motion of falling stones and their effects on talus slope by Machida et al. (1975) in Japan. In an experimental field of Ashio Mountains, Kanto District, the test site was a talus slope of 90 m in length and its slope angle 35° in average inclination. A wooden chute of 1.5 m in length was placed over the talus slope apex keeping its end side 2 m in height over the ground surface and 45° inclination. The diameters of stones used ranged from -3 to $-9\ \phi$ (8 mm to 512 mm) and the numbers of stones tested in each class of grain size (1ϕ-scale interval) were 18 to 50. Immediately after dropping the stone onto the talus slope through the chute, measurements were made of the travel distance of the test stone and its transit time through the measuring sections which were set at intervals of 2-10 m along the slope. The transit time was recorded with an electric pulse on a pen-recorder with manual switching by the observers watching at every measuring section.

Generally a coarser rock fragment can travel a longer distance along the slope, but some deviation from this tendency is found depending upon the sorting state of superficial talus materials. The velocity measurement showed that a smaller particle (larger than -5ϕ) decelerates downslope because of increasing of the ratio of diameter of superficial particles to that of falling stone, while a larger particle (smaller than -5ϕ) accelerates some distance before decelerating and has a peak (maximum) velocity at a point in the falling course. An approximately linear relationship between the peak velocity V (m sec^{-1}) and the median diameter of falling stones d (in ϕ-scale) can be expressed as $V = 1.75d - 4.1$ ($d = \leq -3.5\phi$) with a correlation coefficient of 0.98. But this empirical formula may not apply to other talus slopes because of the difference of superficial materials.

Kotarba (1981) carried out a field experiment of debris transportation and deposition on slopes in the Tatra Mts. during 1975-80. He recognized that geomorphic work on the sheet talus and talus cone is greatest close to the rock wall while in avalanche chutes the maximum was found to occur in the transition zone extending between avalanche niche and trough. Change in accumulation amounts and movement rate of surficial material with distance from the rock wall were measured and described quantitatively in relation to Caine's model (Caine 1969).

As for a modified type of 'fall', toppling failures from alpine cliffs on Ben Lomond, Tasmania were investigated by Caine (1982), who pointed out two different mechanisms – cambering process and slab failure as dominant causes of the toppling through topographic evidence and joint surveys.

3.5.2 *Observation of 'slide' and 'flow' phenomena*

'Slide' and 'flow' are two types of rapid mass movement and there is a difference of moving state between them from a view point of dynamics. But in field work, some common observation methods can be applied to both phenomena. For a field study on rapid mass movement, an adequate selection of observation times is as important as a set of comprehensive observations of many physical factors. A sensor system should be established to detect the occurrence of 'slide' or 'flow' exactly and to send a signal to an observation or recording system to start their operation. For an explanation of such a system, a comprehensive observation system for debris flow is introduced which has been developed by Okuda et al. (1980a) and succeeded in getting useful information about rapid motion of debris flows at Mt. Yakedake. A schematic diagram of this on-line system for observation control is illustrated in Figure 3.12. Figure 3.13 shows the control room instrumentation and a real observation system set at the eastern slope of Mt. Yakedake is shown in Figure 3.14.

This system has the following functions: the arrival of a debris flow front is detected by the multi-sensors located along the valley bottom at some definite intervals (see Fig.

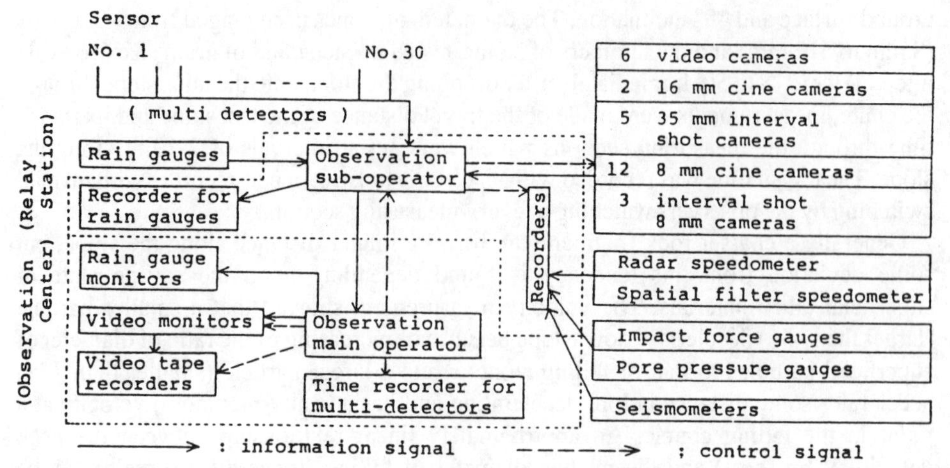

Figure 3.12. Observation control system for debris flow observation at Mt. Yakedake, Central Japan.

Figure 3.13. Mt. Yakedake debris flow observation control room.

3.14) and the information signal of the arrival of a debris flow front at each sensor is sent to the observation center through the relay station and each arrival time detected by each sensor is recorded by the main observation operator which sends out a control signal instantly to various corresponding instruments.

Each sensor, instrument or special equipment is explained in the following:

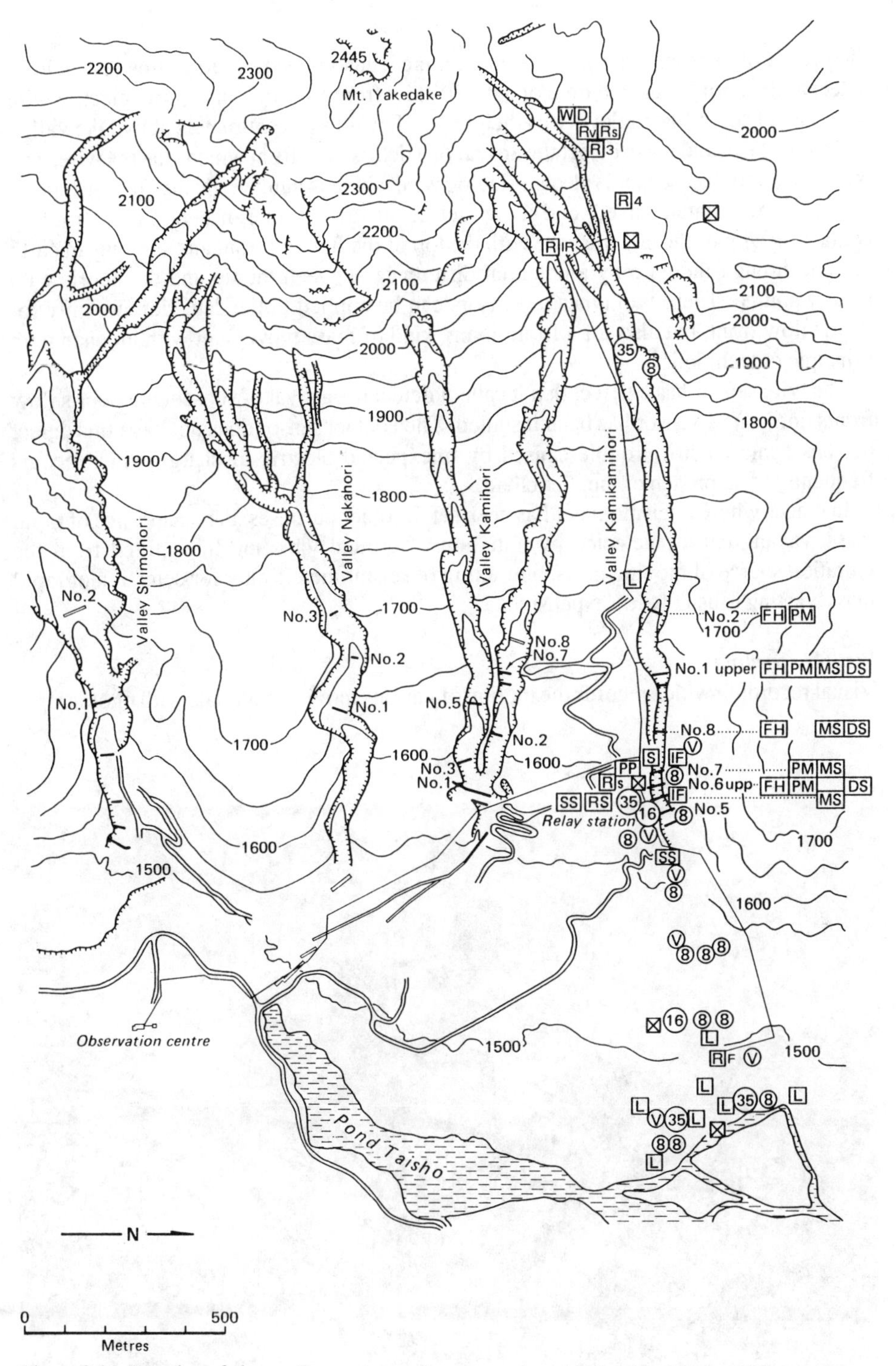

Figure 3.14. Location of observation system at the eastern slope of Mt. Yakedake in 1980.

Sensor

Two types of sensors, wire sensors and contact sensors, were used in this test field to detect and record the arrival time of a debris flow front. A wire sensor is an electric wire which is set across the valley with a height of a few tens of centimeters above the valley bottom surface, and it can detect the arrival of a debris flow front which cuts the wire. The other one, a contact sensor is also an electric wire which is hung vertically from the upper wire rope horizontally fixed, and it can detect the arrival time of debris flow front by contacting a tip of the electric wire with the top of the flowing front and sending an earth current through the wire by a potential difference between the tip and the debris with battery source. These two types of sensors can be applied to the detection not only for debris flow front, but also for a front of any kind of mass movement without change in principle or technique.

The wire sensor has a defect that it cannot detect the arrival of subsequent debris flow fronts until a new sensor has been installed. The contact sensor does not have this defect but has some electric trouble caused by unexpected electric discharge which occurs frequently by approach of thunder cloud.

In places where a seismometer has been set for other purposes, the occurrence of rapid mass wasting can be detected by initiation of ground vibration transmitted from the vibration source of moving mass, but setting of seismometers only for detection of rapid mass wasting is usually too expensive.

Visual recording

Visual recording which records the motion of various moving materials with photograph-

Figure 3.15. Video cameras in field position.

ic camera or videocamera is a useful method for field observation of rapid mass wasting, because it makes it possible to reproduce the motion pattern in time and to analyze the motion quantitatively. But special care has to be taken to apply this visual recording system to field experimental work quite differently from indoor application. For example, a special shading box is necessary for protecting the recording instruments (optical or video cameras) from heavy rainfall, strong wind, direct sunshine, etc. to keep normal functions of the instruments, and periodic inspections are needed to check their functions (Fig. 3.15).

35 mm, 8 mm photographic cameras and cinecameras to shoot at constant intervals appointed by the observer or to record the motion continuously were set along the test valley as shown in Figure 3.14, according to their functions and special purposes for field observation. Some cinecameras were modified to shoot at a constant interval (1 hr, 2 hr, or 1 day) in an ordinary period without rapid mass wasting in order to record the natural state before the occurrence of rapid mass wasting, and worked as a normal cinecamera in an extraordinary period of rapid mass wasting in order to record the moving state of materials.

Videotape recording with video camera has been adopted in the test valley recently because the resolving power of tape images has increased and the costs of the camera and tape have decreased. Videotape recording also usually starts automatically by receiving a signal from the main operator, but sometimes this recording can be operated directly by hand of an observer who is watching a monitor television.

Some standard scales for length should be set in the visual field of the cameras at each recording for quantitative analysis of the recorded images. Usually distinctive targets are set in the visual field with suitable three-dimensional distribution by setting several man-made targets or by selecting fixed points e.g. a large stone which the moving materials cannot remove.

Standard scale for time is also necessary for quantitative analysis of motion by visual recording, so recently an exact recording time can be superimposed on the recording image (photo film or videotape image) automatically with a special attachment for time marker. In case of the cinecamera, relative time interval can be determined from the number of film frames per unit time without any attachment. But the speed is not controlled adequately in normal cameras so that some calibration for film speed with a standard time marker is necessary for strict scientific investigation.

In the work of analyzing the motion of objects from a recorded image, the reading of their exact positions and the following of change over time are important but troublesome. Some special equipment, the so called 'motion analyzer' has been adopted and its usefulness is recognised especially in the field of engineering, for example, the exact pursuit of flying rockets. Of course, this equipment can be used in the quantitative analysis of rapid mass wasting effectively when the scale for distance and time are set sufficiently in the recorded image.

But for usual problems on rapid mass movement, most motions are limited in one or two dimensions and the speeds are not so high, and an approximate treatment of the images without special equipment can give a satisfactory result for study of characteristic motion within an allowable accuracy. For example, a speed of debris flow front can be estimated from successive photographs which had been taken at constant time interval of 1.00 s regulated by a crystal oscillator and showed the front position by target points in the visual field. It is possible to observe surface features of the moving mass and to

Figure 3.16. Spatial filter speedometer (1), radar Doppler effect speedometer (2) and video camera (3) to observe debris flow motion.

measure the size distribution of moving materials larger than 10 cm in diameter by setting camera in an appropriate position.

Direct measurement of speed of mass

As mentioned above, the average speed of the front of mass motion can be measured by the multi-sensor system, while the speed of each particle of moving materials on the surface can be measured by the visual recording method. But the former method can be applied only to the front of the moving mass, and the latter cannot be applied at night unless flood light projectors are available. Other methods to measure the speed of moving mass or moving particles have to be devised independently of the methods using the multi-sensor system or visual recording. Special speedometers based on a noncontacting principle should be adopted, and two kinds of speedometer (a radar speedometer and space filter speedometer) are adopted for the present time in Japan to measure the speed of debris flows. A radar speedometer based on the Doppler effect with ultra-short wave of 10.53 GHZ has been adopted to measure continuously the speed of moving materials through a fixed point in the test valley (as shown in Fig. 3.14 with a special mark RS). In our test field, there is no water flow under ordinary meteorological conditions and a surface runoff or debris flow appears only after a heavy rainfall and continuous measurement is needed only in the period of heavy rainfall. The beam of the emitting wave from the meter has some divergence and the speed is therefore averaged over a surface area.

A second type of speedometer, a so-called spatial filter speedometer (Fig. 3.16), was adopted in 1977 to measure surface velocity of a moving mass continuously and to analyze the variation of speed in time. Some modifications for the special purposes of field observation (Itakura, 1980) were necessary (Fig. 3.14 at the mark SS). This

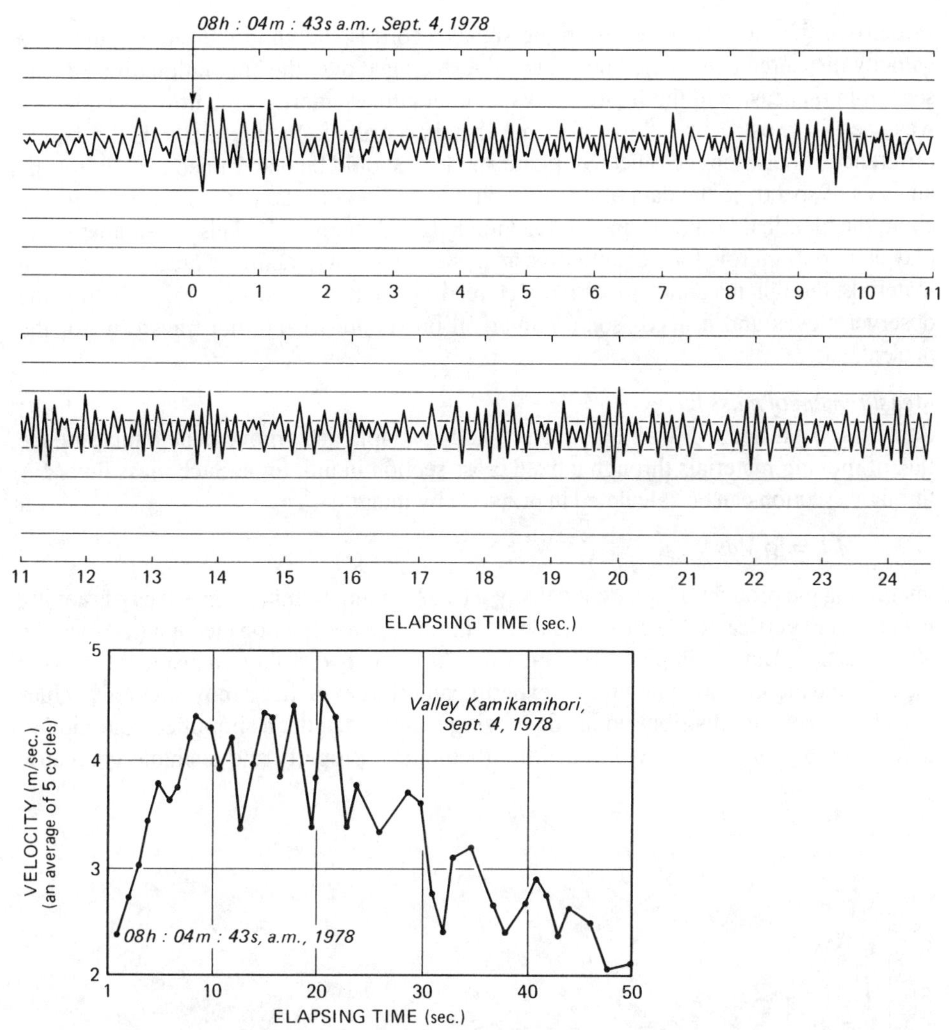

Figure 3.17. Surface velocity of debris flow observed with the spatial filter speedometer on Sept. 14, 1978 at Mt. Yakedake; upper: typical output signal wave for the debris flow; lower: surface velocity obtained from the above signal.

speedometer is composed of a parallel-slit reticle sensor, electric amplifier and data recorder. The reticle sensor can transduce the surface velocity on the objective plane to an angular velocity of optical image through an objective lens and parallel-slit reticle and further transduce the angular velocity to an electric signal through a photocell. From the continuous record of the signal, the velocity of moving materials or its spectrum can be obtained after some numerical treatment. The measurement of surface velocity of debris flows by this speedometer has been carried out since early 1977 and successful results have been obtained on several debris flows. A part of typical wave signal from the sensor responding to a debris flow on Sept. 4, 1978 is shown in Figure 3.17 and the surface velocity of this flow analyzed from the signal record is shown in the same figure. The velocity measured by this method is the value averaged over the integration time i.e. one second in this case and the figure shows an undulating pattern of the flow speed which was seen in most debris flows also by visual recording. In this method, the height difference H between the objective plane and lens should be determined exactly for the analysis of speed, so the data of photographic recording or ultrasonic water level gauge about the identical flow was used to read the height difference H. This speedometer can play an important role for a quantitative analysis of special dynamic character of moving materials though it cannot give a direct reading of the speed of motion before the observer's eyes and it needs some numerical procedure for Fourier transform for the present.

Measurement of mass flux

Sometimes, it is necessary for quantitative study on rapid mass movement to measure the flux of moving materials through a fixed cross section in unit time. Such mass flux $M.F.$ through a section can be calculated in principle by integral

$$M.F. = \int \rho . V. ds$$

integrating the product of apparent density ρ (mass in unit volume), the speed of moving materials (V) vertical to the cross section and the small cross section element (ds) over the whole section. But it is impossible to measure the velocity distribution inside the moving mass. Some theoretical or laboratory experimental studies on mass movement give clues to estimate velocity distribution inside moving materials under a simplified condition of flux of uniform substance with a definite rheological property and a simple geometric

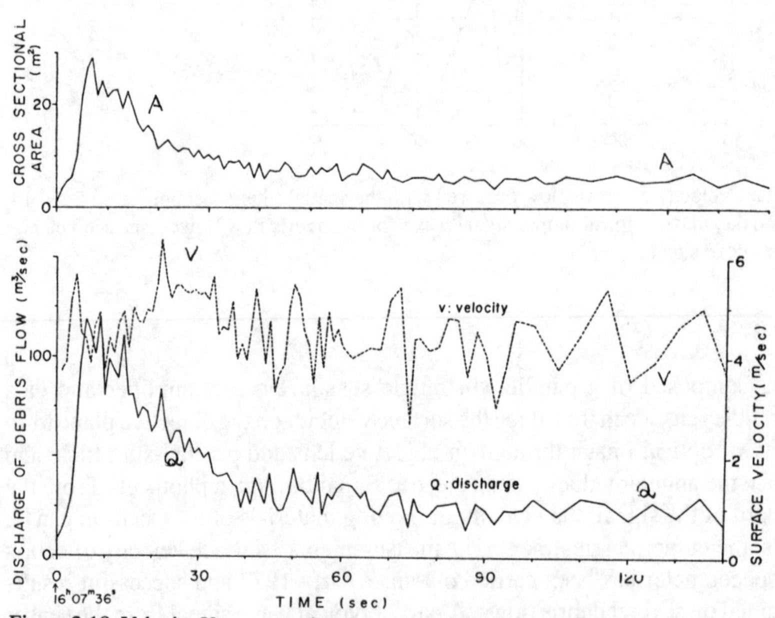

Figure 3.18. Velocity V, cross sectional area A and discharge $Q = VA$ of a debris flow observed on Aug. 23, 1980.

cross section and in such a simple case, a relationship between surface velocity and mean velocity averaged over the whole section can be determined reasonably. But in a real case where the moving materials are not uniform and the cross section takes an irregular shape, the velocity distribution inside the moving mass cannot be estimated.

About a cross sectional area of mass flux, it is very difficult to determine the area because the surface of the moving mass is not flat but usually undulating and the shape of bottom of the mass which may be changeable owing to the scouring or depositional process in the midst of motion cannot be observed by any method.

Still more, an apparent density of moving materials also can be hardly determined in the field because the moving material is not homogeneous through the whole mass depending upon the mixing ratio of solid fragments, water and air, and the sampling of moving matcrials is very difficult as mentioned in the next section.

Therefore, only a rough calculation on mass flux is possible by an approximate estimation of average speed, cross-sectional area and apparent density of moving materials at a specified section as shown in the following example in the test field.

As a cross section convenient for field observation of rapid mass wasting, an open section over a check dam is very suitable for measurements of mass flux because the bottom shape of moving mass is unchangeable and known by the fixed bed surface of concrete or fixed stones and comparatively uniform velocity distribution can appear through the section because of less bottom friction on flat and smooth surface, and the cross sectional area can be calculated by visual recording method owing to its simple geometric shape. Still more, the apparent density also can be measured more easily at this section than at the other section, because setting of equipment for sampling of moving materials is easy in this section as described in the later explanation.

An example of observations of mass flux in the test field is shown in Figure 3.18. This figure shows the change of flow velocity (V) measured by the space filter speedometer, cross sectional area (A) calculated from the photographic recording at time interval of one second, and the volume flux $V \times A = Q$ of a debris flow respectively which passed over the check dam No. 6 in the test field on Aug. 23, 1980. The curves show that the volume flux has a peak immediately after the arrival of debris flow front. The high rate of flux is confined to the initial period of about 30 seconds, the velocity is comparatively uniform

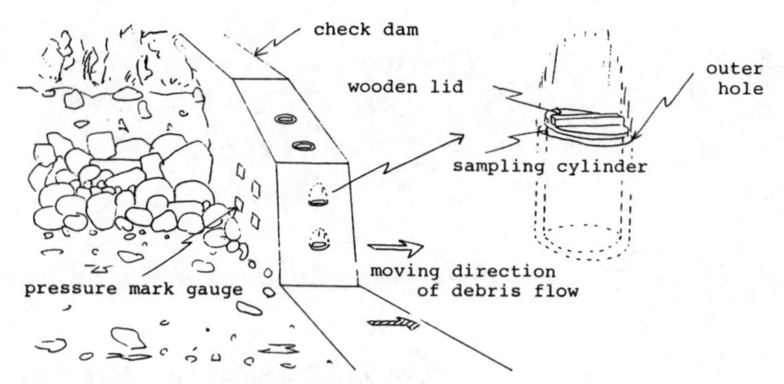

Figure 3.19. A sampling cylinder set on a check dam wall for debris flow materials.

Figure 3.20. Debris sampler in field position.

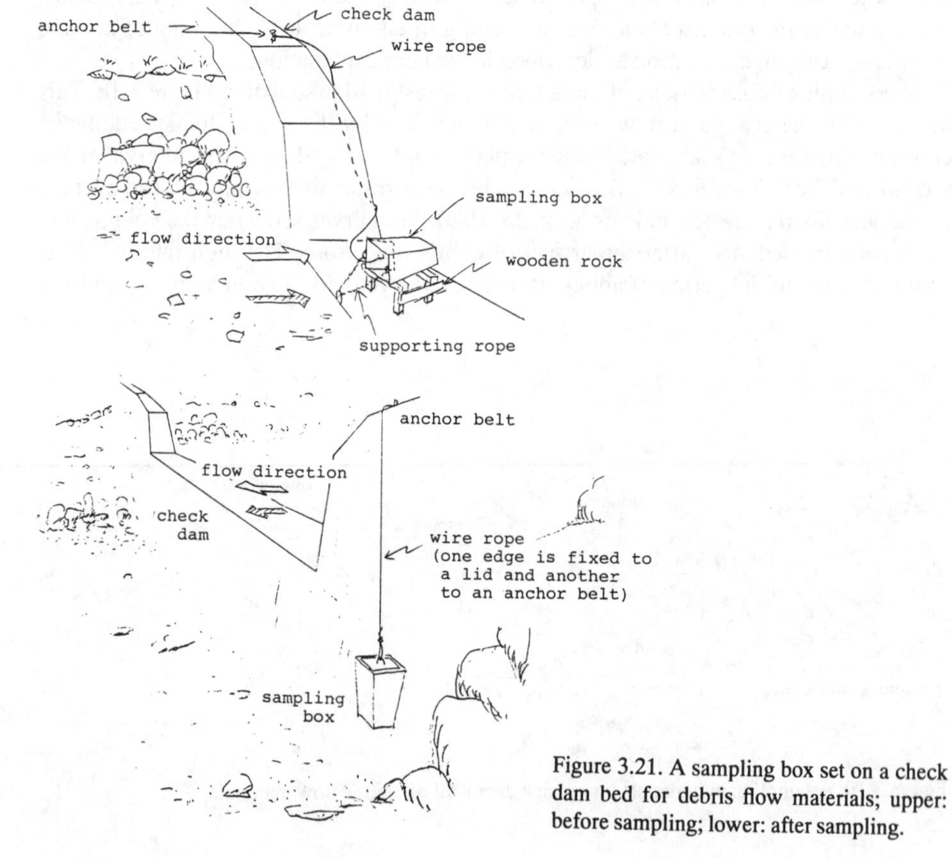

Figure 3.21. A sampling box set on a check dam bed for debris flow materials; upper: before sampling; lower: after sampling.

on average but with wide variations from that average and the cross sectional area has an early peak.

Sampling of moving materials

It is important for the analysis of rapid mass movement to investigate the physical properties of moving materials, so some samples have to be picked up from the moving mass without changing the physical state.

One method we tested was to trap a small lump of moving materials into a hole on the dam. Steel cylindrical bottles (12 cm diameter, 15 cm depth) with a wooden lid were set in the holes which were dug in the concrete walls of check dams as shown in Figure 3.19. The bottles were taken to a laboratory to test various properties of trapped materials. The wooden lid is put over the upper mouth of the cylinder by its weight only and it can be pushed away instantly by the arrival of a moving front. This method is very simple and can catch samples very easily under suitable conditions, but before catching the moving materials of debris flow, a small amount of rain waters have been often trapped in the bottles during a long period without debris flow.

Then, a new method was devised to correct the above defect and to catch a larger volume of moving materials only from the debris flow front. A large sampling box (entrance 30 × 30 cm, length 60 cm) was set horizontally on a supporting rack with a height of 30 cm above the floor of open section of check dam, holding the entrance upward of channel as shown in Figure 3.20. Sands and pebbles transported by usual small runoff cannot enter the box because of clearance 30 cm between the floor and the lower edge of the entrance. When a front arrives at the box entrance, moving materials enter into the box and push back the box along the moving direction cutting down the supporting rope of the box before the entrance. Then the box falls down from the dam floor and hangs along the back side of dam wall by a strong wire fixed to the inner lid of the box as shown in the figure, and it does not catch further materials. By this new method, we have succeeded in catching a sufficient volume of samples for various tests including cobbles with 10 cm diameter (Fig. 3.21).

Needless to say, the bottle or box can catch only the materials from the head arriving first at the setting place and no succeeding samples. But we have to take successive samples from the mass following the head because the composition of materials is remarkably changeable along the longitudinal direction.

3.5.3 *Other observations on dynamic phenomena*

In previous sections, special experimental methods are introduced to get quantitative information about dynamic phenomena. But careful observations can supply valuable knowledge of dynamic phenomena even without special system or equipment if they are carried out taking important dynamic factors into account.

As an old example, in the midst of rockfall and sturzstrom at Elm 1882, a village teacher had observed the motion of collapsed debris with a stop watch in his hand until a part of his house was buried by the debris and he gave a vivid account of the event which was cited by Prof. Albert Heim as the best scientific record (Hsu 1978).

As a recent example, in the midst of a large-scale landslide in Fukuchi, Hyogo Prefecture, Japan, 1978 which was caused by a heavy rainfall (about 500 mm over 2 days) and moved a mass of debris of 6 × 10 m^3, a town officer had taken a series of photographs of the collapsed slope every few minutes ascertaining the shooting time with his watch, and a town inhabitant had pictured the debris motion with a cinecamera. From these visual records, an exact kinematics of large-scale landslides with a small forerunning slide and main slumping failure was studied by Okunishi (1982).

As a useful application of field measurement, successful prediction of transition from creep to slide was performed by Yamada et al. (1970) by measuring carefully the creep rate at a steep slope of Mt. Takabayama, Central Japan where a railway tunnel passed. They found the change in the creep rate curve of the slope and could predict the occurrence time of slope failure a few days in advance of actual failure. Then trains were stopped for safety and the direct damage from the landslide was prevented beforehand.

As for field observations on debris flow, effective field surveys have been executed in various countries. Pierson (1980) observed directly the erosional and depositional processes by debris flows during one storm and obtained valuable information about the nature of debris flows at Mt. Thomas, New Zealand. Kinetic factors (intervals of surges, speed of debris mass and particle size distribution) and geomorphic features (flow levee construction and fanhead trenching) were observed and discussed in relation to causes of the debris flows. Rapp and Nyberg (1981) surveyed the debris flows near Abisko, N. Sweden and recognized more dominant effect by debris flows on the depositional process than by running water, rockfall and slush avalanching in the region. They measured the diameters of existing lichens for the purpose of dating past debris flows and estimated preliminarily the return period of the flows at about 50 to 200 years. Li Jian et al. (1983) had carried out field observation on mudflows in the mountainous area of southwest China over 15 years with manually operated recording system and clarified the static and dynamic features of the mudflows and their depositional process. Suwa and Okuda (1983) tried a trench cut survey of debris flow deposits and recognized the inverse grading structure in relation to the motion of the debris flow.

3.6 LABORATORY WORK, MODELLING AND SIMULATION

Laboratory investigation into rapid mass movement is of four kinds: mapping, geotechnical tests, physical model experiments and theoretical modelling.

3.6.1 *Mapping*

Special care has to be taken on mapping rapid mass movement features. Rapid change in topography can be pursued only by repeated surveys in special areas of large relief and steep slopes. Recently airphotos are often used to make a map of mountainous regions where access is difficult, but it costs very much and the accuracy of height determination is inadequate for most geomorphic purposes. Precise equipment for mapping from stereo photos needs a special operator. But a simple plotter combined with a stereoscope can be operated after short training and is sufficient for tracing topographic changes in a limited area if suitable photographs can be obtained.

A simple map to illustrate extraordinary phenomena appearing on the ground surface e.g. a pattern of cracks before slope failure or a distribution of large debris fragments transported by a debris flow can be plotted by a simpler method. Ordinary photographs taken by ordinary cameras even from a model airplane or a balloon can be used if some standard targets are distributed in the visual field of the camera and their positions are plotted on a large scale map. Sometimes terrestrial stereo photos are utilized effectively as mentioned in Section 3.2.1.

Besides the topographic mapping, digital mapping has developed remarkably and information about the geometry of the ground surface can be stored as digital memory in a computer. Necessary data or some quantitative information derived from the data can be obtained directly from the computer and a figure or a map expressing the spatial result can be drawn with a graphic device connected to the computer. This technique is effective for study of rapid mass movement in which some changes in topography or in physical state must be investigated by comparing successive data in time series, and the results must be shown by a figure or a map.

Recent progress of remote sensing techniques can supply us with valuable information about various characteristics of the ground surface, such as the difference of physical state or plant cover between different slopes.

3.6.2 *Geotechnical tests*

Many geomorphologists are carrying out geotechnical laboratory tests in relation to studies of rapid mass movement. There are many properties of slope materials which must be tested in the laboratory, but the numbers of properties are very limited if a direct relation to mass movement phenomena is emphasized.

One of the most important properties in relation to the cause of mass movement is shear strength which prevents the slope materials from moving downward as described in Section 3.1.2. To calculate strength from in-situ measurements, an empirical relation of laboratory and field measurements is required.

By laboratory tests we can measure directly the strength expressed by internal friction angle (ϕ) and cohesion (c) of sampled materials, and special care has to be taken to keep samples under conditions as close as possible to those in the field though the samples will have been disturbed by sampling and remolding for the test.

There are two common tests used for laboratory measurement (the direct and the triaxial shear test) and three kinds of test procedures are adopted according to the drainage and consolidation conditions (the unconsolidated undrained (UU), the consolidated undrained (CU) and the consolidated drained (CD) test). The best method should be selected to simulate the failure phenomenon under the limited conditions as correctly as possible.

Concerning slope failure, the Mohr's strength theory of failure is used most commonly, and careful consideration is necessary to apply the σ-τ relation which varies depending upon the kinds of slope materials. Many test examples for various practical problems have been published in text books of soil or rock mechanics in which wide experiences are obtained in various field conditions and on various materials (e.g. Bishop and Henkel 1964).

Rodine et al. (1974) performed a synthetic approach for study on the initiation and mobilization of debris flows. They deduced mechanisms of occurrence and motion of debris flows, and devised a new method using paired special conical penetrometers, by which Coulomb's strength of soft and remolded debris could be determined. They explained the high mobility of debris flows taking the effect of mixing of different size particles on the strength into consideration. They also defined a new 'mobility index' which is a measure of the potential for debris flow and can be determined by geotechnical tests in laboratory for the sampled materials. They showed the effectiveness of the new index as a basis for determining a criterion of debris flow initiation based on field data.

Besides the test of shear strength, the permeability of slope materials is often measured in the laboratory to investigate vertical infiltration of water from ground surface or ground water flux parallel to the slope through a permeable layer. This is because the distribution of ground water and especially of its pore pressure as determined by the movement of water has a direct connection with the initiation of slope failure.

The procedures to measure the permeability of sampled materials in the laboratory using special apparatus should be referred to the textbooks of soil mechanics. In practical problems, the information about geohydrological properties are usually needed over an area, so special care has to be taken to distribute sampling points in the area and to check the permeability measured in laboratory by comparing them with the values obtained by field tests e.g. pumping or tracing tests for ground water survey.

3.6.3 *Model experiments*

As for examples of fundamental experiments to examine an individual process and to derive a physical law, Takahashi (1977) carried out a hydraulic experiment of debris flow to investigate the fluid dynamical character of grain flow and he showed that velocity distribution in a model debris flow can be derived from Bagnold's dispersive stress theory. Suwa et al. (1984) are performing an experiment on inverse grading in debris flow using a vibrating box to check the effect of kinetic sieving on segregation of different size particles.

Such physical experiments are effective in clarifying the fundamental process quantitatively by controlling a few dominant factors, but cannot give directly any information about synthetic effects of various processes which are working simultaneously and producing actual topographic changes.

As for an example of scale modelling on rapid mass movement, Hsu (1975) carried out a model experiment on catastrophic debris stream (sturzstroms) to reproduce the motion of the famous Elm event in 1881 in which a large rockfall of 10^7 m^3 in volume occurred and the mass of debris ran down about 2 km in 40 s. In his experiment, he determined scale factor of geometry, time, velocity and acceleration based on a similitude law and selected bentonite suspension as a flowing material after trial and error tests. He succeeded in simulating the prototype debris stream by a small model flow in a flume.

Some geomorphological experiments have been carried out to investigate combined effects of several processes which produce a topographic change. Takahashi (1980) made hydraulic experiments on settlement of debris flows on a fan using a mixture of sands with different diameters and changing the mixing ratio of water and sands to investigate the mechanism of fan evolution. From his experiment, he obtained some useful results to predict a dangerous region of debris flows in a fan, and to plan countermeasures for preventing the damage of debris flows.

As for the geomorphological experiments on fan evolution, Hooke (1967, 1979) performed experiments to investigate the mechanism of development of alluvial fans in a test area of 10 m × 10 m discharging a mixture of water and sands. From his experiments, it was found that not only one discharge of debris flow, corresponding to one rapid mass movement, but also a time sequential occurrence of deposition by debris flows and scouring by flood contribute to the development of an alluvial fan over long periods.

Generally, all types of experiments are planned under somewhat simplified conditions and a similitude law is not always established in the model experiments. Therefore, careful check of the difference between model and prototype is needed to interpret experimental results and observations in the field.

3.6.4 *Theoretical modelling and numerical simulation*

A theoretical model based on the results of field and laboratory experiments or tests, and simulation of various geomorphological processes numerically by using suitable values for parameters in the applied model, constitute standard procedure for modern geoscience. Sometimes a theoretical model has been constructed by pure deduction, and the applicability of the model to real phenomena is verified afterwards by field observations or laboratory experiments. An object of theoretical models may be a simple individual process or a synthetic effect of multiple processes which produce a topographic change, as in the case of model experiments.

As a model for accumulation of falling stones, Statham (1976) showed a new model for scree slope development based on rockfall and noticed the characteristic features – a basal concavity of profile, a straight slope angle less than the repose angle and sorting into fine particles at the top and coarse at the base.

Slope stability theory is based on a kind of theoretical model for slope failure which assumes a slip surface in the ground and compares a driving force (or moment) on the mass above the surface and a resistant force (or moment) along the surface. Two types of theoretical models assuming planar or rotational slip surface are most commonly used to investigate the critical condition of initiation of sliding along an infinite plane parallel to the slope or along a circular cylindrical surface in the ground. In the theoretical models, failure in ground materials is analysed according to Mohr's strength theory. Numerical estimation of stability (calculation of safety factor) can be performed by shape of ground materials above the assumed slip surface, bulk density of the mass, the strength of materials (c, ϕ) and the pore pressure along the slip surface which are determined by field observation, in-situ or laboratory tests or reasonable estimation.

Some external impacts may play the role of trigger for slope failure initiation. Earthquakes bring about an additional large force proportional to its acceleration. A heavy rainfall or a rapid slow melt event supplies water into slope materials and raises pore pressure, so its effect also should be estimated in the slope failure model.

High speed computers have made the analysis of slope stability possible under more complex conditions than in the case of simple planar or rotational slip. Slope failures with any shape of slip surface or a transient state of stability caused by water infiltration can be simulated numerically if necessary parameters are given by reasonable estimation or various tests. Classic theory of slope failure should be referred to the text books in soil mechanics. Modern theoretical models including rheological failure are introduced in the technical papers in geomechanics.

As for a theoretical model of debris flow, some modification of a slope stability model with planar slip surface is needed to explain the maintenance of high mobility of moving mass for a long path. Takahashi (1977) constructed a model of debris flow initiation taking the water supply from surface flow to debris mass after fracture into consideration, and he tested the applicability of his model to debris flows by flume experiments by using sands as movable materials and verified the appropriateness of the model.

Still more, Takahashi (1977) applied Bagnold's dispersive stress theory to construction of a grain flow model, as a new theoretical model of debris flow motion. He also showed the applicability of the model to the flow of a mixture of water and sands by a flume experiment, which verified that the velocity distribution in the experimental flows agreed with the one derived from the model.

As another model of debris flows Johnson (1970) used a Bingham flow model and he showed that velocity distributions of small debris flows observed in an experimental flume and a small stream agreed well with the distribution derived from a Bingham model with plug flow. He also showed that the Bingham flow such as debris flow or glacier flow through a valley can form U-shape valleys by special scouring and depositional actions.

Though a contradiction may be found between two models, i.e. grain flow and Bingham flow for the same debris flow phenomenon, the rheological property of a mixture of water and various debris fragments varies widely depending upon the composition of materials.

As for a theoretical model of the depositional process of debris flows, a stochastic 'random walk model' was applied to the prediction of depositional patterns on a fan by Price (1974), and, after some modification, by Imamura and Sugita (1980). The random walk model was applied to prediction of the depositional area of mudflows from Usu volcano, Hokkaido, Japan and good agreement was found between the results from the model using suitable parameters and the deposits of the alluvial fan.

3.7 RAPID MASS MOVEMENT AND HILLSLOPE EVOLUTION

As described in Section 3.1, rapid mass movement usually brings about a considerable change in topography in a short time, so its effect on topographic change is an important problem in geomorphology. The area affected by rapid mass movement varies in a wide range from 10 × 10 m to 1000 × 1000 m depending upon the volume of transported debris mass.

In the uppermost region of a slope or a valley, slope failure usually cuts down a gully along the slope surface or pushes the valley head backward by scouring and changes of drainage pattern.

In the middle region of a slope or a valley, rapid movement of a large volume of debris mass produces a new channel or enlarges a cross section of an existing channel.

Along a lower slope or valley and at a slope foot or valley outlet, settlement of a large amount of debris fragments brings about a thick accumulation of debris. A fluvial channel system in the depositional region is disturbed and changed in a short period by local deposition and scouring. Such a topographic change is propagated gradually from the slope foot or valley outlet to a lower region through a fluvial system over a longer period.

As for an example of topographic change caused by rapid mass movement, quantitative observations along the Kamikamihori Valley in Figure 3.14, where debris flows frequently occur, have been carried out by Okuda et al. (1980b). In their work, valley head retreat (several tens of meters during 15 years), change in valley bed level (a few meters every year) and horizontal progress of fan edge (three hundred meters during 16 years) were calculated from repeated topographic surveys or comparison of airphotos taken successively almost every year. The valley bed level commonly changes a few tens of meters in altitude with the passing of only one large debris flow.

A remarkable topographic change which affects a wide area is occasionally brought about by the damming up of a river with a large amount of debris. A sudden damming up of a river raises the water level in the newly produced reservoir in a short time. If the new natural dam is broken down, a rapid discharge of stored water brings about a flood or serious erosion along the lower reaches, while if the natural dam is kept for a long time, the change in base level of erosion at upper and lower reaches of the dam causes some gradual topographic change along the lower river and the upper reservoir over a long period.

Rapid mass movement can produce various kinds and scales of topographic changes causing such disruption that quantitative studies on the relationship between rapid mass movement and topographic changes are needed from both view points of geomorphological and environmental sciences.

General description of slope development by erosional, transportational and depositional processes can be found in text books of geomorphology and a synthetic review of the historical study of slope profile change from the classical models of W. M. Davis and W. Penck to modern theoretical studies is introduced by Carson and Kirkby (1972).

The individual effect of rapid mass movement on slope development is also explained by Carson and Kirkby (1972). They have described various kinds of effect which are brought about by rapid mass movement and have shown their characteristic features depending upon the properties of slope materials and action of external agents.

The most general principle governing ground surface changes of level is the law of mass conservation, or a continuity equation for debris movement. The law is expressed by the following partial differential equation:

$$\frac{\partial y}{\partial t} = -\frac{1}{\rho}\nabla.\vec{S}$$

where $\frac{\partial y}{\partial t}$ is the change rate of ground surface level y through time t, ρ is bulk density of debris deposits after settling on the slope surface, ∇ means vector 'divergence' and $\vec{S}$ means the flux vector of debris transport (Kirkby 1971).

For simplification, if the problem is limited to two dimensions, e.g. the change of slope profile in a cross section of a long mountain range, the equation is simplified as in the following:

$$\rho\frac{\partial y}{\partial t} = -\frac{\partial S}{\partial x}$$

which means that the change of slope surface level corresponds to the difference of debris flux between inflow from the upper side and outflow to the lower side in a small slope surface element with horizontal length Δx as shown in Figure 3.22.

In a simple three dimensional problem, in which debris mass runs down across a curved contour with the radius of curvature R (R is positive when the contour is concave outward), the equation is expressed in the following:

$$\frac{\partial y}{\partial t} = -\frac{\partial S}{\partial x} + \frac{S}{R}$$

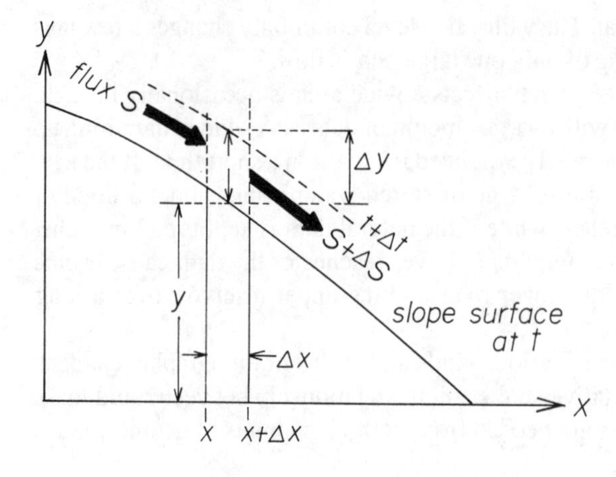

Figure 3.22. Conceptual representation of 'mass conservation' in the case of mass movement along a slope.

Besides the continuity equation as mentioned above, a transportation law should be established to describe the form of the flux S (the value of S is the amount of mass transported through unit cross section width in unit time) and to calculate the slope development caused by debris movement.

As a general functional form of the flux S, not related to mass movement directly, Kirkby (1971) set up a special condition (transport limited removal)

$$S = ax^m \left(\frac{\partial y}{\partial x}\right)^n$$

He solved the two dimensional equation approximately by using various combinations of values of m and n and showed a dimensionless graph of the characteristic slope profile for a range of processes $0 < m < 2.5$ and $1 < n < 2$.

As another functional model of S derived from physical meaning, Hirano (1968), set the complex form of S

$$S = (\alpha + \beta x)\left(\frac{\partial y}{\partial x}\right)$$

where term $\alpha\left(\frac{\partial y}{\partial x}\right)$ corresponds to creep and $\beta x\left(\frac{\partial y}{\partial x}\right)$ to wash respectively. He showed that this complex model can explain the slope change better than any other single model, and solved analytically the equation under various initial and boundary conditions.

Apart from general expression of flux S, a suitable functional or numerical expression of S is first needed to estimate quantitatively the hillslope evolution caused by rapid mass movement, but as shown in Section 3.5, it is difficult for us to get quantitative information about the flux of debris transport in the midst of a rapid mass movement phenomenon except in the specified observation field with special equipment. Even though the form of flux S is determined exactly by direct observation in specified time and place, it is also difficult to obtain the reliable solution through an integrating procedure over a long period and a wide slope by using the determined value S, because some factors controlling the flux are changeable in time and space. For example, a climatic change may occur over a long period and different slope materials show different properties across a wide slope.

Therefore, it seems to be impossible to predict quantitatively the tendency of hillslope evolution from some results of field observation for the present. However, as for the short-term change of specific slope caused by the rapid mass movement, it can be estimated roughly if field experimental works including careful observation have been prosecuted there suitably for the special type of mass movement. For an example, in Kamikamihori Valley in Mt. Yakedake, Central Japan, retreat rate of valley walls, scouring rate of valley bed and depositional rate of debris fragments at the fan have been evaluated by field experimental works on debris flows about the average, maximum and minimum states during fifteen years, and the change in slope along the valley and the fan can be estimated within a probable range. But, a possible large change over a wide slope of Mt. Yakedake which may be caused by volcanic eruption at intervals of several tens of years cannot be predicted at present stage because reliable data have not been accumulated about the eruption and its aftereffect on mass movement.

Apparently contrary to the above mentioned difficulty, a long-term prediction of hillslope evolution in relation to mass movement can be prosecuted under special conditions such as when the assumption of uniform functional form of debris flux (S) is acceptable as averaged over the long period and wide area.

As an example of such a successful procedure, Hirano (1972) tried to explain the slope development of fault scarp of Mt. Hira, west coast of Lake Biwa, where rapid mass movement is proceeding by rockfall and landslide, by using his model. The results obtained from morphometry of the fault scarp showed a good agreement with his calculated one using suitable parameters, and difference of erodibility between granite and Palaeozic sediments is expressed by differential value of erosional coefficient.

This result means that a theoretical model can describe the hillslope evolution by rapid mass movement if suitable average parameters are used for the flux (S) even though no information about individual mass movement has been obtained.

As another example of hillslope evolution in relation to mass movement, Mizutani (1970) calculated the dissection rate of the mountain body of Yotei Volcano in Hokkaido, Japan and its hillslope development according to his own model taking debris transport by water flow and mass wasting into consideration. The result of his calculation accounted for the actual dissecting process.

For comparison between calculated results of hillslope evolution and actual slope, it is important to check the absolute time interval between two stages – i.e. present and a definite past time. A dating technique, suitable for special problems, has to be selected.

At the present time, ^{14}C method, lichenometry, weathering-rind thickness and dendrochronology are often utilized to date the age of some special samples which were carried in or buried under ground by rapid mass movement. Development of convenient and reliable methods of dating are required for experimental geomorphology.

Special care is needed to study a catastrophic effect of large-scale landslide (slide volume > 10^6 m^3) on slope development, because it can change suddenly even whole features of a mountain and can leave the large topographic effects for a long time over wide area and it often brings about secondary mass movement of smaller scale for a long period after the largest mass movement.

REFERENCES

American Society of Civil Engineers 1975. *In Situ Measurement of Soil Properties.*

Bagnold, R.A. 1957. The flow of cohesionless grains in fluid. *Trans. Royal Soc. London*, 249A, 235-297.

Bishop, A.W. and Henkel, D.J. 1964. *The measurement of soil properties in the triaxial test*, 2nd ed. Edward Arnold Ltd.

Caine, N. 1969. A model for alpine talus slopes development by slush avalanching. *J. Geol.*, 77, 92-100.

Caine, N. 1982. Toppling failures from alpine cliffs on Ben Lomond, Tasmania. *Earth Surface Processes and Landforms*, 7, 133-152.

Carson, M.A. and Kirkby, M.J. 1972. *Hillslope form and process.* Cambridge University Press.

Churchill, R.R. 1979. A field technique for profiling precipitous slopes. *British Geomorphological Research Group Technical Bull.* 24, 29-34.

Gardner, J.S. 1983. Rockfall frequency and distribution in the Highwood Pass area, Canadian Rocky Mountains. *Z. Geomorph.*, 27, 311-324.

Hills, R.C. 1970. The determination of infiltration capacity of field soils using the cylinder infiltrometer. *British Geomorphological Research Group Technical Bull.*, 3, 1-25.

Hirano, M. 1968. Mathematical model of slope development. *J. Geosciences*, Osaka City University, 11, 2, 13-52.

Hirano, M. 1972. Quantitative morphometry of fault scarp with reference to the Hira Mountains, Central Japan. *Japanese J. Geol. and Geog.*, 17, 85-100.

Hooke, R.L. 1967. Processes on arid region alluvial fans. *J. Geol.*, 75, 438-460.

Hooke, R.L. 1979. Geometry of alluvial fans: effect of discharge and sediment size. *Earth Surface Processes*, 4, No. 2, 147-166.

Hsu, K.J. 1975. Catastrophic debris stream generated by rock falls. *Geol. Soc. Amer. Bull.*, 86, 1, 129-140.

Hsu, K.J. 1978. Albert Heim, Observations on landslides and relevance to modern interpretations. *Rockslides and avalanches: Developments in geotechnical engineering* 14A, Elsevier, 71-93.

Hutchinson, J.N. 1982. Methods of locating slip surfaces in landslides. *British Geomorphological Research Group Technical Bull.* 30, 1-30.

Iida, T. and Okunishi, K. 1979. On the slope development caused by the surface landslides. *Geographical Review of Japan*, 58, 2, 426-438 (in Japanese with English abstract).

Imamura, R. and Sugita, M. 1980. Study on the simulation of debris depositing based on a random walk model. *Shin-Sabo*, 32, 3, 17-26 (in Japanese).

Ishii, T. 1981. Microforms and slope processes of the Ashio mountains in Central Japan. *Trans. Japanese Geomorphological Union*, 2, 279-290.

Itakura, Y. 1980. Debris-flow-velocity sensor with a parallel-slit reticle. Explanatory pamphlet for excursion of *IGU commission on field experiments in geomorphology*, Japan.

Johnson, A.M. 1970. Rheology of debris, ice, lava. *Physical processes in geology*, 14, Freeman, Cooper and Company.

Kirkby, M.J. 1971. Hillslope process-response models based on the continuity equation. *'Slopes, form and process'*. Institute of British Geographers, 15-30.

Kotarba, A. 1981. Present-day transformation of Alpine granite slopes in the Polish Tatra Mts. *Trans. Japanese Geomorphological Union*, 2, 179-186.

Li Jian et al. 1983. The main feature of the mudflow in Jiang-Jia Ravine. *Z. Geomorph.* 27, 325-341.

Luckman, B.H. 1976. Rockfalls and rockfall inventory data: some observations from Surprise Valley, Jasper National Park, Canada. *Earth Surface Processes*, 1, 287-298.

Machida et al. 1975. Formation processes of a talus cone in the Ashio waste land. *Geographical Review of Japan*, 48, No. 11, 768-783 (in Japanese with English abstract).

Miller, J.P. and Leopold, L.B. 1963. Simple measurements of morphological changes in river channels and hillslopes. *Changes of climate*, UNESCO Arid Zone Research Series XX, 421-427.

Milton, B.D. 1976. *Introduction to geophysical prospecting.* 3rd ed. McGraw-Hill.

Mizutani, T. 1970. Erosional process of Yotei Strato-Volcano in Hokkaido, Japan. *Geographical Review of Japan*, 43, 32-44.

Okimura, T. 1983. Rapid mass movement and groundwater level movement. *Z. Geomorph*. Supp. Bd. 46, 35-54.

Okuda, S. et al. 1976. Synthetic observation on debris flow (part 2) observation in 1975. *Disaster Prevention Research Institute Annals* 19, B-1, 385-402 (in Japanese with English synopsis).

Okuda, S. et al. 1978. Synthetic observation on debris flow (part 4) observation in 1977. *Disaster Prevention Research Institute Annals* 21, No. 1, 277-296 (in Japanese with English synopsis).

Okuda, S. et al. 1979. Synthetic observation on debris flow (part 5) observation in 1978. *Disaster Prevention Research Institute Annals* 22, No. 1, 157-204 (in Japanese with English synopsis).

Okuda, S. et al. 1980a. Observation of debris flow at Kamikamihori Valley of Mt. Yakedake. *Excursion guidebook for 3rd meeting of IGU commission on field experiments in geomorphology*, 116-139.

Okuda, S. et al. 1980b. Observation on the motion of a debris flow and its geomorphological effects. *Z. Geomorph. Supp*. 35, 142-163.

Okunishi, K. 1982. Kinematics of large-scale landslides – a case study in Fukuchi, Hyogo Prefecture, Japan. *Trans. Japanese Geomorphological Union*, 3, 41-56.

Pierson, T.C. 1980. Erosion and deposition by debris flows at Mt. Thomas, North Canterbury, New Zealand. *Earth Surface Processes*, 5, 227-247.

Price, F.W. 1974. Simulation of alluvial fan deposition by a random walk model. *Water Resources Research*, 10, No. 2, 263-274.

Rapp, A. 1960. Recent development of mountain slopes in Karkevagge and surroundings Northern Scandinavia. *Geografiska Annaler*, 42, 73-200.

Rapp, A. and Nyberg, R. 1981. Alpine debris flows triggered by a violent rainstorm on June 23, 1979. *Trans. Japanese Geomorphological Union*, 2, 329-341.

Rodine, J.O. et al. 1974. *Analysis of the mobilization of debris flows*. NTIS. AD/A-001832.

Sakanoue, T. 1969. Study on precipitation in mountainous region. *J. Faculty Agric.*, Kyushu University, 24, No. 1, 29-113.

Sato, T. et al. 1950. Some measurements by new type of mountain infiltrometer (1st report). *Report of Forestry Experiment Station*, Japan, 83, 39-64.

Statham, I. 1976. A scree slope rockfall model. *Earth Surface Processes*, 1, 43-62.

Suwa, H. et al. 1983. Topographic changes on the sidewall and in the valley bottom of the Kamikamihori Valley of Mt. Yakedake. *Disaster Prevention Research Institute Annals* 26, No. 1, 413-435 (in Japanese with English Synopsis).

Suwa, H. et al. 1984. Size segregation of solid particles in debris flows, Part 1. *Disaster Prevention Research Institute Annals* 27, No. 1, in press (in Japanese with English synopsis).

Suwa, H. and Okuda, S. 1982 Sedimentary structure of debris-flow deposits, at Kamikamihori fan of Mt. Yakedake. *Disaster Prevention Research Institute Annals* 25, No. 1, 307-321 (in Japanese with English synopsis).

Suwa, H. and Okuda, S. 1983. Sedimentary structure of debris-flow deposits, at Kamikamihori fan of Mt. Yakedake. *Disaster Prevention Research Institute Annals* 26, No. 1, 307-321 (in Japanese with English synopsis).

Takahashi, T. 1977. A mechanism of occurrence of mud-debris flows and their characteristics in motion. *Disaster Prevention Research Institute Annals* 20, No. 2, 405-435 (in Japanese with English synopsis).

Takahashi, T. 1980. Study on the deposition of debris flow (2) – Process of formation of debris fan. *Disaster Prevention Research Institute Annals* 23, No. 2, 443-456 (in Japanese with English synopsis).

Tanaka, S. and Okimura, T. 1976. A relationship between soil structure and slope failure depth in natural slopes. *13th Symp. natural disaster science 1976 at Kyoto*, 237-238 (in Japanese).

Yamada, G. et al. 1970. Collapse of Takabayama tunnel, Iiyama-line caused by a landslip. *Report of Railway Technical Research*, JNR. No. 706 (in Japanese).

Yokoyama, K. 1983. Private communication to the author.

Zaruba, Q. and Mencl, V. 1982. *Landslides and their control*. Czechoslovak Academy of Sciences, 2nd ed.

Shimizu, [illegible] 19[illegible]. Experimental process of [illegible] in Hokkaido, Japan. [illegible]

[illegible] 19[illegible]. [illegible] mass movement [illegible] 20: 2[illegible]

[illegible], S. et al. 1976. [illegible] observations in 1974 [illegible] 16, 17-23. (In Japanese with English abstract)

[illegible], et al. 1978. [illegible] observations of 1977. [illegible] (In Japanese with English abstract)

[illegible] et al. [illegible] observations in 19[illegible] 32, No. 2, 137-[illegible] (In Japanese with English abstract)

[illegible] Valley [illegible] M. V. [illegible]

[illegible] 19[illegible] [illegible]

[illegible] 1982. [illegible] Research, [illegible]

[illegible] C. 19[illegible]. [illegible] North Carolina. [illegible] Zealand [illegible] 17

[illegible] 19[illegible]. [illegible]

[illegible] 19[illegible]. [illegible]

[illegible] and [illegible] 19[illegible]. [illegible]

[illegible] 19[illegible]. [illegible] 1977 [illegible]

[illegible] 19[illegible]. [illegible]

[illegible] 19[illegible]. [illegible] 68

[illegible] 19[illegible]. [illegible]

[illegible] 19[illegible]. [illegible] (In Japanese with English abstract)

[illegible] 19[illegible]. [illegible] (In Japanese with English abstract)

[illegible] (In Japanese with English abstract)

[illegible] 19[illegible]. [illegible] (In Japanese with English abstract)

[illegible] 19[illegible]. [illegible] (In Japanese with English abstract)

[illegible] and [illegible] 19[illegible]. A relationship between [illegible] (In Japanese)

[illegible] 19[illegible]. [illegible] (In Japanese)

[illegible] 19[illegible]. [illegible]

[illegible] 19[illegible]. [illegible]

CHAPTER 4

Surface wash

RORKE B. BRYAN
University of Toronto, Canada

4.1 INTRODUCTION

Surface erosion processes are rarely precisely defined and some initial clarification is therefore essential. Surface wash, which is approximately synonymous with sheetwash and sheet erosion, describes the removal of soil from extended surface areas without apparent concentration along drainage lines. The results are clearly distinct from rill or gully erosion, but the distinction between processes is much less obvious. Sheet erosion was initially envisaged as the removal of an essentially uniform layer of soil. However, this would imply uniformity of both tractive force application and surface resistance. This is apparently precluded by the extremely variable depth of surface wash, and the demonstrated heterogeneity of most surface materials. Most surface wash therefore contains concentrated streams of deeper, more rapidly flowing water which is hydraulically indistinguishable from rill flow. Much of the erosion associated with surface wash probably occurs in such concentrated zones so that distinction from rill erosion becomes largely theoretical. Rill wash is probably a more accurate description of the situation on most hillslopes.

In view of the interaction of processes, discussion of surface wash must also concern rill initiation. Although sheet and channel processes interact intimately in most field situations, small variations result in distinct surface forms which reflect particularly the definition and preservation of rill channels. In some cases, although erosion is active, variations in the balance of tractive force and surface resistance are not sufficient to produce defined channels. In other cases such channels form but are eliminated during storms by processes such as rainsplash on interrill areas or sedimentation during flow recession. Sometimes channels do survive after flow but are eliminated by short term or seasonal influences affecting surface materials. In other circumstances channels may persist as semi-permanent features which eventually grow into gullies or perennial stream channels. It is desirable that experiments on surface wash and rill initiation be sufficiently prolonged to allow the long-term persistence of rills to be assessed.

Under most circumstances the onset of runoff and surface wash during a storm is preceded by a period during which rainsplash alone is active. In intense rainstorms the soil surface may undergo great disturbance and crusting which may accelerate the initiation of runoff. When runoff starts particle entrainment may result from flow drag and lift forces alone. However, interaction of rainsplash and sheetwash is usually very

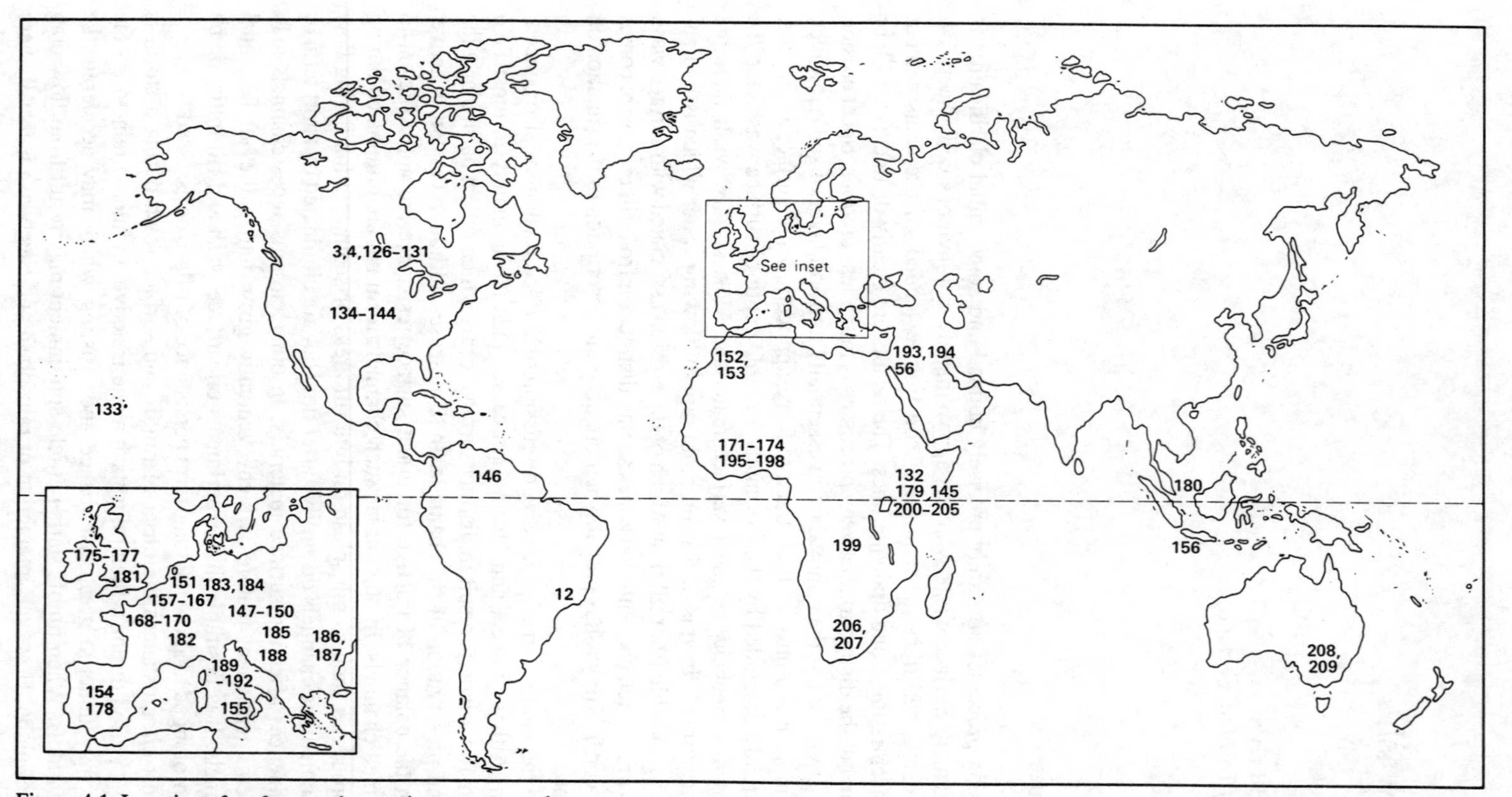

Figure 4.1. Location of surface wash experiments reported.

effective with entrainment by splash and transport by wash in rainwash erosion (De Ploey et al. 1976). The diameter of particles on the surface often equals or exceeds the flow depth, further complicating hydraulic patterns and ensuring that particles which do move are transported as saltating and rolling bedload rather than suspended load. Unless the flow is unusually deep, raindrop impact will also directly affect many particles so that erosion can be accurately characterized as runoff creep or rain creep (De Ploey and Moeyersons 1975).

Disturbance of flow patterns and the soil surface by rainsplash is intense during the early stages of surface wash while flow depths are small. This produces highly irregular flow and turbid mixture of sediment and water. On certain soils under intense rainfall as moisture contents approach the liquid limit, or as positive pore pressures are induced by raindrop impact, liquefaction occurs adding shallow mass movement to the mixture of processes. Eventually increasing depths of surface water absorb raindrop energy and protect underlying surfaces, allowing flow to become more orderly and reducing turbidity levels. As flow typically increases in depth downslope, such protection usually starts near the slope base, though surface microtopography and vegetation may result in localized, irregular patterns on many hillslopes. Protection will gradually extend upslope, eventually transforming surface wash into a form essentially unaffected by rainsplash. Although this is the theoretical end point of any rainstorm, intense rainfall is seldom sufficiently prolonged to see it realized. On most hillslopes, therefore, when surface wash occurs it consists of a highly disturbed, irregular and turbid interaction of rainsplash and surface flow.

The origins of field experimental research on surface wash processes extend back at least to the pioneer work of Ewald Wollny in Germany in the late nineteenth century (Wollny 1895). Since then thousands of experiments have been carried out by researchers from many disciplines, but particularly from soil conservation, agricultural engineering, hydraulic engineering, hydrology, forestry and geomorphology. These experiments defy easy classification either by method or objective but very broadly they may be grouped into the following categories:

a) runoff plot experiments designed to isolate soil erosion processes and major contributory factors;

b) runoff plot experiments designed to test the impact of cropping, tillage or soil conservation practices on soil erosion;

c) runoff plot or hillslope experiments focussed on hydrologic processes, particularly infiltration, lateral sub-surface flow and patterns of runoff generation;

d) runoff plot or microcatchment experiments designed to identify threshold rainfall conditions for runoff infiltration and sediment entrainment on different surfaces, as components of catchment water and sediment budget studies;

e) experiments carried out to identify denudation rates as part of denudation chronology studies;

f) experiments designed to isolate surface wash processes as components in hillslope evolution studies;

g) runoff plot and microcatchment experiments focussed on the hydraulic characteristics of surface wash and its relationship to rill and channel initiation;

h) experiments related to morphometry and the development of drainage networks.

These are not mutually exclusive and many experiments have contributed results in more than one category. Many different methods have been employed, reflecting differing objectives, local availability and cost of equipment and the conceptual frameworks employed by researchers. Although inevitable, this is unfortunate as it does seriously limit the comparative use of data.

All the categories of study identified have been based primarily on field research. It should be emphasized, however, that many aspects of surface wash processes are extremely difficult to study with precision in the field. As a result, many researchers have also incorporated laboratory experiments and some of the most significant progress has resulted from such work. Unlike laboratory studies of, for example, fluvial processes, most of these experiments are at, or close to, full scale and the results can therefore be applied to the field situation without scaling difficulties.

It is not possible in the space available to review even a small sampling of the vast volume of published material resulting from early experiments. These have been reviewed in a series of excellent recent books, such as Kirkby (1978), Kirkby and Morgan (1980), De Boodt and Gabriels (1980), Morgan (1981) and Zachar (1982). The last-named book is of particular interest as it reviews a large body of work carried out in the Soviet Union and in Eastern Europe which is not well-known in western countries. Reference should also be made to the valuable bibliographies produced by Streumann and Richter (1966) and Richter (1977). This chapter is not intended to repeat these comprehensive reviews, but to provide a sample of recent research on surface wash with a brief description of objectives, methodology and progress achieved, and some suggestions for future research directions. It is, by necessity, incomplete, being based on a limited response to a questionnaire survey, personal correspondence, some visits to experimental installations and a review of some recent literature. It includes some ongoing research and some which, though completed and published, may not yet be widely known. In using published work a cut-off date of 1976 has generally been applied though some reference has been made to a few particularly significant earlier works.

The listing of current or recent experiments related to surface wash shown in Table 4.1 is by no means exhaustive, but it is believed to represent a good cross-section of work in progress in different countries (Fig. 4.1). It is clear that the experiments fall into several distinct groups based on methodology employed and on objectives. One conclusion which must be drawn in review is that there is comparatively little interaction between researchers involved in different types of experiments and little attempt to integrate results. While many aspects of the surface wash processes are unknown or poorly understood and much further experimentation is clearly necessary, a serious attempt to integrate different approaches is an urgent necessity.

4.2 SURFACE WASH EXPERIMENTS BASED ON RUNOFF PLOTS AND PRACTICAL LAND MANAGEMENT PROBLEMS

4.2.1 *Origins in USA*

The earliest origins of erosion research involved confined runoff plot studies and this is clearly still one of the most widely-used approaches. By far the largest body of data available from runoff plots comes from the experimental stations of the US Soil Conservation Service established during the 1930's. The runoff plots used are of standard

Table 4.1. Recent field experiments in surface wash.

Study No.	Country	Researchers	Location	Major objectives a	b	c	d	e	f	g	h	Selected references
126	Canada	Bryan	Toronto	x							x	Bryan (1979, 1981)
3	Canada	Bryan, Hodges	Alberta	x		x	x	x	x	x	x	Hodges & Bryan (1982), Bryan & Hodges (1984), Hodges (1982, 1984)
4	Canada	Bryan, Campbell, Imeson	Alberta	x		x		x	x			Bryan & Campbell (1980, 1982, 1984)
127	Canada	Bryan, Luk	Toronto	x								Bryan & Luk (1981)
128	Canada	Hamilton	Toronto	x		x	x					Hamilton (1984)
129	Canada	Morgan	Toronto	x	x	x			x			Morgan (1979)
130	Canada	Luk	Toronto	x			x			x		Luk (1982, 1983), Luk & Morgan (1981)
131	Canada	Bryan, Bowyer-Bower	Alberta								x	Bowyer-Bower & Bryan (1984)
132	Canada	Bryan, Bowyer-Bower	Baringo	x		x	x	x	x	x	x	
133	USA	El-Swaify	Hawaii	x	x	x						El-Swaify & Dangler (1976)
134	USA	Emmett	Wyoming						x	x	x	Emmett (1970, 1978)
135	USA	Lyles	Kansas	x				x				Lyles (1976)
136	USA	Band	California				x	x				Band (1983b)
137	USA	Ponce, Hawkins	Price River, Utah				x					Ponce & Hawkins (1978)
138	USA	Laronne	Colorado, Utah					x				Laronne (1981, 1982)
139	USA	Wells, Gutierrez	Chaco, Ah-shi-le-pah, New Mexico	x		x	x	x	x		x	Wells & Gutierrez (1982)
140	USA	Parker, Schumm	Fort Collins, Colorado			x	x				x	Parker & Schumm (1982)
141	USA	Karcz, Kersey	Binghampton, New York							x	x	Karcz & Kersey (1980)
142	USA	Yoon, Wenzel	Colorado							x	x	Yoon & Wenzel (1973)
143	USA	Kilinc, Richardson	Colorado							x	x	Kilinc & Richardson (1973)
144	USA	Shen, Li	Colorado							x	x	
145	USA	Dunne, Dietrich	Amboseli			x		x	x	x	x	Dunne & Dietrich (1980, 1980b)
146	Venezuela	Pla Sentis	Northern Plains, Venezuela	x	x	x						Pla Sentis (1980)
12	Brazil	Netto	Cachoera River, Tijuca National	x		x		x				Netto (1983)
147	Poland	Gil, Welc	Flysch Carpathians (Szymbark)	x	x	x						

Table 4.1 (continued).

Study No.	Country	Researchers	Location	Major objectives a	b	c	d	e	f	g	h	Selected references
148	Poland	Slupik	Flysch Carpathians (Homerka and Bystzanka Catchment)			x	x					Slupik (1981)
149	Poland	Froehlich	Flysch Carpathians (Homerka Catchment)			x	x					Froehlich (1982)
150	Poland	Gerlach	Not specified	x								Gerlach (1979)
151	Netherlands	Epema, Riezebos	Utrecht	x							x	Epema & Riezebos (1983)
152	Netherlands	Imeson, Kwaad	Rif Mts, Morocco	x		x						Imeson (1983)
153	Netherlands	Imeson, Kwaad Verstraten	Beni-Boufrah, Morocco	x		x			x		x	Imeson et al. (1982)
154	Netherlands	Imeson	nr Grenada, Spain	x		x	x	x	x	x		Van der Linden (1983)
155	Netherlands	Van Asch	Calabria, Italy	x	x	x		x	x			Van Asch (1980)
156	Netherlands	Van der Linden	Serayu Valley, Central Java	x		x	x	x				Van der Linden (1983)
157	Belgium	De Ploey, Savat, Morgan	Leuven	x						x	x	De Ploey et al. (1976)
158	Belgium	Savat, De Ploey	Leuven			x			x	x	x	Savat (1976, 1977, 1980), Savat & De Ploey (1982)
159	Belgium	De Ploey, Mucher	Leuven	x		x		x	x			De Ploey (1977, 1979, 1981), De Ploey & Mucher (1981)
160	Belgium	De Ploey, Gervers	Leuven (Huldenberg)	x	x	x		x	x		x	
161	Belgium	Poesen, Savat	Leuven							x	x	Poesen & Savat (1980)
162	Belgium	Savat, Poesen	Leuven									Savat & Poesen (1981)
163	Belgium	Poesen	Leuven			x		x	x	x	x	Poesen (1981, 1983a, 1983b)
164	Belgium	Moeyersons	Tervuren					x				Moeyersons (1983), Bollinne & Rousseau (1978)
165	Belgium	Bollinne, Binard, Rousseau	Liege, Hesbaye	x	x			x	x			Bollinne (1978, 1980, 1982), Binard & Bollinne (1980)
166	Belgium	De Ploey, Bryan	Leuven, Toronto	x		x			x			Bryan & De Ploey (1978)
167	Belgium	Gabriels, Pauwels	Ghent	x	x						x	Gabriels et al. (1976)

Table 4.1 (continued).

Study No.	Country	Researchers	Location	Major objectives								Selected references
				a	b	c	d	e	f	g	h	
168	France	Roels	Ardeche, France	x	x	x				x		
169	France	Morand	Mont de Vaux, Cessieres-Aisnes	x		x						Morand (1979)
170	France	Messer	Alsace	x	x							Messer (1978, 1980)
171	France	Valentin	Cote d'Ivoi-re	x	x	x						Valentin (1979, 1981)
172	France	Valentin, Roose	Cote d'Ivoi-re	x	x							Valentine & Roose (1981)
173	France	Roose	Cote d'Ivoi-re, Upper Volta, Benin, Niger	x	x	x						Roose (1981)
174	France	Collinet, Valentin	Upper Vol-ta, Niger	x		x						Collinet & Valentin (1982)
175	United Kingdom	Rowntree	Glasgow			x			x		x	Rowntree (1982)
176	United Kingdom	Morgan, Noble	Silsoe	x		x			x	x	x	Morgan (1978, 1980, 1981, 1982a, 1983b), Noble & Morgan (1983)
177	United Kingdom	Quansah	Silsoe	x	x			x	x	x		Quansah (1982)
178	United Kingdom	Scoging, Thornes	S.E. Spain			x		x	x			Scoging (1980, 1982), Scoging & Thornes (1980)
179	United Kingdom	Rowntree, Ngahu	Baringo, Kenya				x	x	x		x	
180	United Kingdom	Hatch	Sarawak	x	x	x						Hatch (1981)
181	Ireland	Johnston, El-sawy, Cochrane	Six Mile Water Catchment, Northern Ireland			x	x					Johnston et al. (1980)
182	Switzerland	Leser, Schmidt, Seiler, Rohrer	Rheinfelden, Wallbach, Mohlinerfeld, Basel	x	x	x					x	Schmidt (1982, 1979) Leser (1980), Leser et al. (1981)
183	W.Germany	Richter, Negen-dank	Trier, Moselle Valley	x	x	x		x	x			Richter & Negendank (1979), Richter (1978), Richter (1980)
184	W.Germany	Bechner, Schwert-mann Vogel, Auerswald	Bavaria	x	x	x						Bechner (1981)
185	Hungary	Kerenyi, Pinczes	Tokaj, Hungary	x	x	x						Pinczes (1982), Kerenyi & Pinczes (1979)

Table 4.1 (continued).

Study No.	Country	Researchers	Location	Major objectives a	b	c	d	e	f	g	h	Selected references
186	Romania	Dedui, Cazangui	Bucharest	x	x		x					Dedui & Cazangui (1983)
187	Romania	Ionita, Papa, Motoc	Perieni-Birlad	x	x	x	x					Ionita (1983)
188	Yugoslavia	Djorovic	Srbja, Yugoslavia	x	x	x						Djorovic (1980)
189	Italy	Giordano, Mattioli	Sicily, Tuscany, Lombardy, Sardinia	x	x							Giordano & Mattioli (1981)
190	Italy	Chisci, D'Egidio, Zanchi	Vicarello (Pisa)	x	x	x						Chisci & Zanchi (1981), Chisci et al. (1981)
191	Italy	Chisci, D'Egidio	Era Valley (Pisa)	x	x		x					Chisci & D'Egidio (1981)
192	Italy	Zanchi, Bazzoffi, D'Egidio, Nistri, Sfalanga	Era Valley (Pisa)	x	x	x						Zanchi et al. (1981), Bazzoffi et al. (1980), Torri et al. (1980)
193	Israel	Yair, Lavee	Sinai			x	x	x	x		x	Yair & Lavee (1981), Yair & Ruttin (1981), Yair et al. (1980a), Yair & Danin (1980)
194	Israel	Yair, Bryan, Lavee, Adar	Zin Badlands	x		x	x	x	x		x	Yair et al. (1980b)
56	Israel	Yair	Negev	x	x	x		x	x		x	Yair (1983)
195	Nigeria	Lal, Aina, Taylor	Ibadan	x	x	xx						Lal (1967a-f)
196	Nigeria	Lal	Ibadan	x	x	x	x					Lal (1982), Lal (1983)
197	Nigeria	Lal, De Vleeschauwer, De Boodt	Ibadan	x	x							De Vleeschauwer et al. (1978)
198	Nigeria	Lal, Bryan, Imeson	Okumo-Udo (nr Benin City)	x	x	x	x			x	x	
199	Zaire	Soyer, Miti, Aloni	Lubumbashi, Shaba	x	x							Soyer et al. (1982)
200	Kenya	Thomas, Tefera	Kabate (Nairobi)	x	x	x						Tefera (1984)
201	Kenya	Ulsaker	Katumani (Machakos)	x	x	x						Ulsaker (1982)
202	Kenya	Thomas, Barber, Moore	Machakos, Kabete, Baringo	x	x	x						Barber et al. (1981), Barber et al. (1979), Moore et al. (1979), Barber et al. (1983)
203	Kenya	Aubrey, Wahome	Amboseli	x	x	x				x	x	Othieno (1979)
204	Kenya	Othieno	Kericho	x	x	x						
205	Kenya	Smith, Critchley, Chekwony	Baringo		x	x	x					Smith & Critchley (1983)
206	South Africa	Gerland	Umfolozi Game Reserve	x	x							

Table 4.1 (continued).

Study No.	Country	Researchers	Location	Major objectives a	b	c	d	e	f	g	h	Selected references
207	South Africa	Platford	Mount Edgecombe, Natal	x	x							Platford (1979, 1982), Platford & Nel (1980)
208	Australia	Reeve, Perrens		x								Reeve (1982)
209	Australia	Moss, Green, Hutka, Walker	Canberra	x						x	x	Moss et al. (1982, 1980)

Major objectives:

a. Runoff plot experiments designed to isolate soil erosion processes and major contributory factors.

b. Runoff plot experiments designed to test the impact of cropping, tillage or soil conservation practices on soil erosion.

c. Runoff plot or hillslope experiments focussed on hydrologic processes, particularly infiltration, lateral sub-surface flow and patterns of runoff generation.

d. Runoff plot or microcatchment experiments designed to identify threshold rainfall conditions for runoff initiation and sediment entrainment on different surfaces, as components of catchment water and sediment budget studies.

e. Experiments carried out to identify denudation rates.

f. Experiments designed to isolate surface wash processes as components in hillslope evolution studies.

g. Runoff plot and microcatchment experiments focussed on the hydraulic characteristics of surface wash and its relationship to rill and channel initiation.

h. Experiments related to morphometry and the development of drainage networks.

dimensions (22.1 m long × 1.85 m wide), standard rectangular shape, and are located on gently sloping agricultural land, typically of about 5° inclination. These plots have produced an immense body of data concerning the impact of different tillage practices and crop types on soil loss and runoff production. This was employed by Smith and Wischmeier (1965) to develop the Universal Soil-Loss Equation and its subsequent modification (Smith and Wischmeier 1978).

Although the USSCS experiments and the USLE approach have been of great practical value in soil erosion control in the United States their empirical nature and the absence of clear linkage with physical process has greatly limited their applicability in other countries. As a result much of the more recent research in other countries has been established to develop comparable data which can be used to adapt the USLE approach. Some of the most important work has been carried out in tropical regions stimulated by the immediate urgency of soil erosion. In Africa the earliest runoff plot experiments (e.g. Staples 1934, Van Rensburg 1955, Hudson 1957) were contemporaneous with those in the United States, and ultimately generated a different soil erosion prediction model (Elwell 1977).

4.2.2 *Experiments in Africa and the tropics*

Many of the most significant surface wash experiments using runoff plots have been carried out in Africa, and one notes particularly the elaborate and prolonged experiments carried out by Roose and his associates from ORSTOM over a twenty-year period in

Table 4.2. Runoff (in % of annual or daily rainfall) and erosion (t/ha/year) in West Africa from naturally vegetated, bare and cultivated ground (after Roose 1976, 1981, Lal 1976 a-f).

Stations	Slope	Mean annual runoff (%)			Maximum daily runoff (%)			Erosion (t/ha/yr)		
	(%)	Natural	Bare soil	Cultivated	Natural	Bare soil	Cultivated	Natural	Bare soil	Cultivated
Adiopodoumé: ORSTOM, 1954-1975										
'Forêt dense'	4.5	–	35	–	–	98	–	–	60	–
R = 2100 mm	7	0.1	33	0.5-30	0.7	95	60-87	0.017	138	0.1-100
Ferrallitic soil on sand	11	0.3	–	–	1.3	–	–	0.034	–	–
	20	0.5	24	–	3.2	76	–	0.052	570	–
	65	1.2	–	–	12	–	–	0.455		
Anguededou: ORSTOM, 1966-69	29	–	–	0.3-1 (md = 0.5)			2-4 (md = 2.2)			0.06-3 (md = 0.2)
Azaguie: ORSTOM, 1966-74										
'Forêt dense'	14	0.4-5.5	–	5.5-12	–			–		
R = 1750 mm		(md = 2)		(md = 9)	3-39	–	25-74	0.05-1.4	–	0.7-4.5
Ferrallitic soil on chlorite schist					(md = 14)		(md = 60)	(md = 0.15)		(md = 1.8)
Divo: ORSTOM, 1967-74)										
'Forêt dense'	10	0.5-1.4	–	0.3-0.4	3-6	–	1-2.4	0.1-0.6	–	0.06-0.1
R = 1450 mm										
Ferrallitic soil on granite										
Bouabé: ORSTOM, 1960-73										
'Savane arbustive dense'		Pro. 0.03	15-49	0.1-26	Pro. 0.2	40-70	5-65	Pro. 0.001	11-52	0.1-26
R = 1200 mm	4	FP. 0.23			FP 2			FP 0.050		
Ferrallitic soil on granite										
Korhogo: ORSTOM, 1967-74										
'Savane arbustive claire dégradée'	3	FP. 1-5	25-40		8-30	67-89	–	0.01-0.160	3-9	–
R = 1400 mm		(md = 3.0)	(md = 33)		(md = 19)			(md = 0.110)	(md = 4)	
Ferrallitic soil on granite										
Saria: ORSTOM, 1971-74										
'Savane arborée claire à épineux'	0.7	Pro. 5-8	35-43	10-37	Pro. 20-30	69-71	40-65	Pro. 0.2-0.7	14-35	3-14
R = 850 mm	1.7	Pat. 10			Pat. 41			Pat. 0.17	(mo=20)	(mo = 7.3)
Ferruginous soil on shallow cuirasse		Pro. 0.4			Pro. 1-8			Pro. 0.10		

Gonsé/Gampela: ORSTOM, 1967-74										
'Savane arborée claire à épineux'	0.4	Pro. 0.2			Pro. 1			Pro. 0.02-0.05		
R = 850 mm		FP 2.5			FP 8-10			FP 0.05-0.15		
Ferruginous tropical indurated soil		FT 15			FT 50-70			FT 0.41		
at depth	0.8				–	–	50-70	–	–	–
on surface				2-45	–	–	50-70	–	10-21	0.6-10
Séfa (Senegal): ORSTOM, 1954-63										
'Forêt sèche claire'	1-2	Pro. 0.1-1.2	25-55	8-10	–	–	–	Pro. 0.02-0.2		
R = 1300 mm		FP 0.3-1.5						FP 0.02-0.5	30-55	2-20
Ferruginous tropical washed soil on granite										
Agonkaney (Dahomey): ORSTOM, 1964-69										
'Fourré dense'	4.4	0.1-0.9	AR 17	20-35	2.5	69	70-80	0.3-1.2	AR 17-28	10-85
R = 1300 mm										
Ferrallitic soil on sand										
Ibadan (Nigeria): IITA 1972-73										
'Savane arbustive dense'	1		31-58	0.1-15	–	70-89	20-40		5-10	0-1.6
R = 1200 mm	5		38-62	3.2-36	–	70-100	40-50		43-156	0.1-11
Ferrallitic soil on granite	10		36-37	3.4-26	–	80-94	40-70		59-233	0.1-7
	15		30-57	2.9-25	–	70-88	30-60		116-229	0.1-43

Pro = completely protected, md = median, Pat = open range pasture, FP = early fire, mo = arithmetic mean, R = rainfall, FT = recent fire, AR = after reclamation.

Niger, Cote d'Ivoire, Upper Volta and Dahomey (Roose 1981, Valentin 1979, Collinet and Valentin 1982). The experiments included a variety of plot sizes and lengths, but many were of similar dimensions to those used in the United States. The American experiments were based on natural rainfall, but those of Roose and his associates made extensive use of simulated rainfall as well as natural rainfall, thus greatly shortening the necessary experimental period, and permitting testing of a wide range of slope, crop and natural vegetation conditions. The simulator constructed at Adiopodoume was of the Swanson rotating boom pattern (also used by Platford in Natal) and was capable of sprinkling two 10 × 5 m plots simultaneously at intensities of 30, 60, 90 and 120 mm h^{-1}. The kinetic energy produced by the simulated rainfall was approximately the same as natural rainfall for the 90 and 120 mm h^{-1} intensities and 1.2-1.5 times greater than that of natural rainfall.

Some of the major results of the experiments are summarized in Table 4.2. They permit evaluation of all the components of the USLE for use in West Africa and form the basis of numerous practical recommendations for modifying agricultural practices to reduce erosion by surface wash. In evaluating the significance of contributary factors Roose and his associates found that rainfall erosivity was by far the most critical factor, ranging up to 40 times the level recorded in Mediterranean areas. In accordance with this it was found that agricultural practice, and particularly crop cover, was vitally important in controlling surface wash, providing for a potential thousand-fold variation in resulting soil-loss (Table 4.3). On gentle slopes of the old African shield the role of surface wash was found to be limited to transport of particles entrained by rainsplash or detached by animals, but on steeper slopes it split into channels and became capable of entrainment itself.

Table 4.3. Vegetation × cultivation technique (C factor) factors for different cultivation practices in West Africa (from Roose 1981).

	C annuel moyen
Sol nu	1
Forêt, fourré dense, culture bien paillée	0.001
Savane et prairie en bon état	0.01
Savane ou prairie brûlée ou surpâturée	0.1
Plante de couverture à développement lent ou plantation tardive, première année	0.3 à 0.8
Plante de couverture à développement rapide ou plantation hâtive, première année	0.01 à 0.1
Plante de couverture à développement lent ou plantation tardive, deuxième année	0.01 à 0.1
Maîs, mil, sorgho (en fonction des rendements)	0.4 à 0.9
Riz de plateau en culture intensive	0.1 à 0.2
Coton, tabac en deuxième cycle	0.5 à 0.7
Arachide (en fonction du rendement et de la date de plantation)	0.4 à 0.8
Manioc, première année et igname (en fonction de la date de plantation)	0.2 à 0.8
Palmier, hévéa, café, cacao avec plantes de couverture	0.001 à 0.3
Ananas à plat (en fonction de la pente) plantation hâtive	0.001 à 0.3*
avec résidus brûlés	–0.2 à 0.5
avec résidus enfouis } Plantation tardive	0.1 à 0.3
avec résidus en surface	0.001 à 0.01
Ananas sur billons cloisonnés (pente 7%), plantation tardive	0.1

*Valentin & Roose (1981).

The major focus of Roose's experiments was the relationship between major controlling variables of the erosion process and surface wash, but observations were also made of the behaviour of tropical ferruginous and ferallitic soils under intense rainfall. These showed that the most critical soil property was the vulnerability to surface modification and crusting which strongly influenced the timing and quantity of surface wash. This aspect was examined in greater detail by Valentin (1982) with micromorphologic soil analysis. This work, carried out partly with the 'Swanson' type simulator and partly with a sprinkling infiltrometer on 1 m^2 plots (Asseline and Valentin 1978) was combined in a study of the effects of crusting on infiltration rates (Collinet and Valentin 1982).

Although the experiments described were intended primarily for development of practical land use measures, they were used also for initial geomorphic analysis and form an important geomorphic data base. They have been paralleled by another important set of experiments for agricultural purposes carried out by Lal and his colleagues at the International Institute of Tropical Agriculture, Ibadan, Nigeria. These started in 1972 with the clearance of forest from ferruginous alfisols, and many are still in progress. The initial experiments involved twenty 25 × 4 m experimental runoff plots on rectilinear slopes of 0.5, 2.5 and 7.5° (Fig. 4.2). Four additional plots were established on irregular slopes ranging from about 5 to nearly 10° in steepness, with lengths ranging from 12.5 to 37.5 m. Apart from standard flow recording and sediment collection instrumentation, each plot was equipped with access tubes for neutron moisture probes. Over the period of operation the experiments have provided a very large body of data on the influence of different tillage, mulching, weeding, fertilizing and cropping patterns on runoff generation, soil-loss and nutrient entrainment under natural rainfall. Detailed descriptions of the

Figure 4.2. Runoff plot installations of R. Lal at the International Institute of Tropical Agriculture Ibadan, Nigeria.

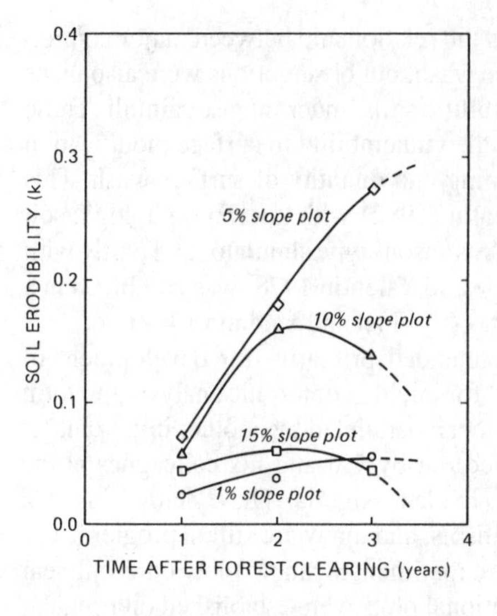

Figure 4.3. Changes in the soil erodibility (K) factor with time after forest clearing (from Lal 1976).

experimental installations are given in Lal (1976a), and results are described in numerous subsequent publications (e.g. Lal 1976b-f, Lal et al. 1979, Lal 1980). Apart from the central purpose of assessing optimal agricultural practices, the experiments have also been used in a wide range of related studies. These include the energy environments and erosivity of tropical rainfall, the influence of organic litter on soil microfauna and their effect on aggregation properties and soil erodibility, the effect of slope length on runoff and soil-loss (Lal 1982), and studies of the effect of forest clearance on soil properties and erodibility (Aina et al. 1977) (Fig. 4.3).

The initial experiments at IITA have been supplemented by a number of additional experiments. These include experimental establishment of relative soil erodibility with a laboratory rainfall simulator (De Vleeschauwer et al. 1978) and a series of runoff plot studies. These involve examination of the effect of plot shape on runoff and soil-loss through small plots of equal area and differing shape, a series of plots (Fig. 4.4) under traditional agriculture, and a set of very large plots set in a cleared catchment of approximately 25 ha area. The last study is of particular interest as it links runoff plot experiments with monitoring of instrumented catchments. Fourteen 200 × 50 m plots are arranged around the catchment in replicate pairs in which various agricultural practices and terrace systems can be evaluated. Each plot is equipped with a large H-flume and recording instruments, but runoff and sediment transport data are also collected in the central catchment channel. The catchment experiment was started in 1979 and some data have been published (Lal 1982, 1983). A similar experiment is currently being set up in the high rainfall belt of southern Nigeria at Okumo-Udo (near Benin City) where natural rainforest is currently being cleared for oil palm production.

A number of plot experiments have been carried out in northern Tanzania to evaluate agricultural practices and soil erosion, particularly on steeply sloping coffee lands near Kilimanjaro (Temple 1972). The important work of Hudson in Zimbabwe has also been noted, and this was followed by a series of soil erosion studies by Elwell and Stocking some of which were related to surface wash and some of which are still in progress. In

Figure 4.4. Instrumented catchment and large runoff plots with diverse land-use practices at IITA, Ibadan, Nigeria.

Tanzania a significant contribution was made by the joint Dar-es-Salaam/Uppsala Universities Soil Erosion Project between 1968 and 1972. This concerned a wide variety of land-use and water management problems, but many of the studies were distinctly geomorphic and hydrologic in nature. They included some measurements of denudation

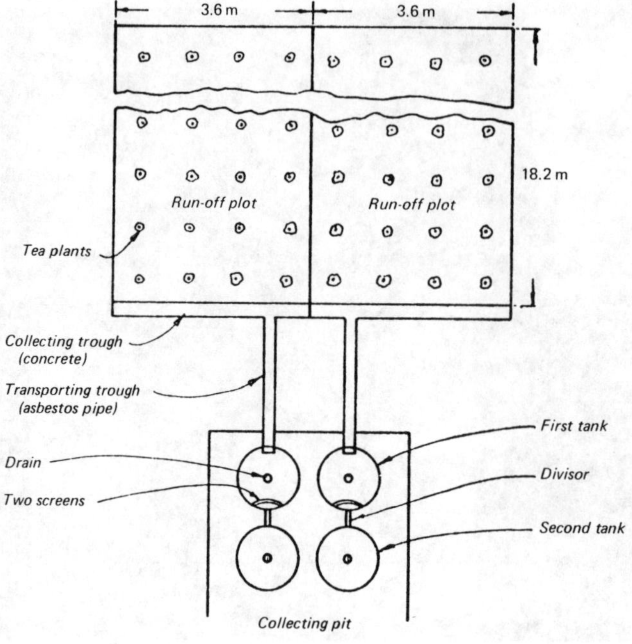

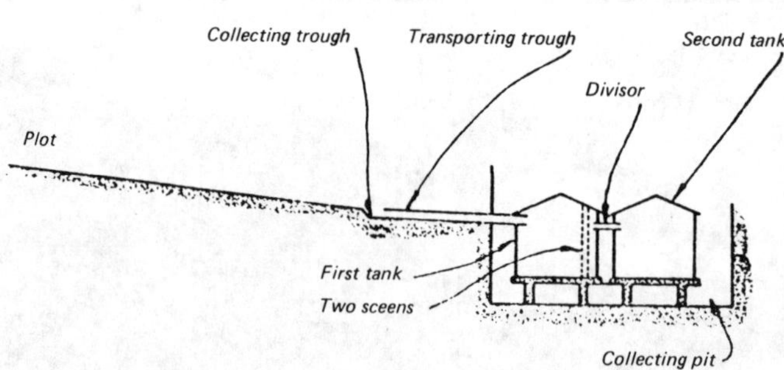

Figure 4.5. Design of runoff plots of C.O.Othieno in the experimental catchment at the Tea Research Institute, Kericho, Kenya (from Othieno 1979).

by sheetwash under varied land use practice in the Uluguru Mountains. Measurements were made by surveyed stakes on plots which were 10 m wide and varied in length from 98 to 142 m (Temple and Murray-Rust 1972). Numerous observations of sheet wash were made in connection with sediment and water budgets of a series of catchments (Rapp et al. 1972), but there were few detailed measurements of sheetwash processes. However, it was concluded that sheetwash is the most significant erosional process in the areas studied.

In Kenya effort has been put into the elaborate and important EAAFRO catchment experiments in which the impact of different plantation and land-use practices on sediment and water yield (Pereira et al. 1962, Blackie et al. 1979) was examined. Othieno (1979) set up sixteen 18.3 × 3.7 m runoff plots at Kericho near the Western Mau Forest Reserve. These were operated under natural rainfall with a water and sediment collection system essentially similar to that of Hudson (1957), (Fig. 4.5). The objective was evaluation of the hydrological effect of four different cropping and tillage practices associated with young tea plants. The results showed a dramatic reduction in soil erosion (approximately one-hundred fold) over a three year period, despite comparable rainfall. This was attributed to the time-dependent development of surface crusts and was interpreted as evidence of the great significance of rainsplash in generating entrainment by sheetwash.

Another series of runoff plot experiments was carried out by Thomas, Barber and Moore (Barber et al. 1979, 1981, Barber 1983) in the Machakos District and at Kabete near Nairobi. These were based on small (1.5 m^2) runoff plots under rainfall simulated by the rotating-disc pattern rainfall simulator designed by Morin et al. (1967) in Israel. Plots were established on 6° slopes and subjected to ½, 1 and 2 h storms at intensities of 25, 50 and 100 mm h^{-1}, each storm being simulated in dry and wet antecedent moisture conditions. Higher soil loss and runoff was recorded from the Machakos luvisol than the Kabete nitosol under all storm conditions, this behaviour being closely related to soil structural conditions and to surface seal formation on the luvisol. These experiments have been followed at Kabete by current experiments with twelve larger runoff plots (12 × 2 m) on a 4.5° slope under simulated and natural rainfall (Tefera 1984).

More recently runoff experiments have been conducted at Baringo in the northern Rift Valley as part of the Baringo Pilot Semi-Arid Area Project. Research on surface wash at Baringo is currently being expanded with several new experimental studies. These include experiments on processes of micro hillslope evolution on different lithologies under simulated rainfall (Rowntree and Ngahu) and detailed studies of sheetwash and rill initiation under simulated rainfall as part of hydrologic and sediment transport budget studies in an instrumented catchment (Bryan and Bowyer-Bower). These experimental studies are more clearly geomorphic in focus than those described above, but will still be closely linked to practical land management measures.

In addition to the studies described another important set of experiments has been carried out in Amboseli National Park in Southern Kenya by Dunne and Dietrich (1980a, b). These studies of surface wash processes and flow hydraulics under simulated rainfall are more clearly geomorphic in focus than any of the results yet published from Kenya and will be discussed in detail in a later section.

Comparatively few reports are available of surface wash experiments with runoff plots in the tropics outside Africa. In Brazil Netto has initiated experiments with simulated rainfall on small undisturbed plots under natural forest. The initial results presented (Netto 1983) show that surface wash occurs primarily as litter flow within the litter cover, and is strongly influenced by the composition and structure of the litter. In Venezuela, Pla Sentis (1980) examined the erodibility and infiltration characteristics of four soils on small plots (1 × 1 m or 3 × 1 m) on slopes ranging from 0.5 to 2.5°. Soils were tested under rainfall intensities of 100 mm h^{-1}. Runoff generation was found to be closely related to crusting behaviour or a 'sealing index'. Extended tests were carried out to assess the impact of treating soils with asphalt mulch. This prevented sealing and greatly enhanced infiltration, with drastically reduced runoff and soil erosion.

Lal (1977) has provided a review of earlier erosion research in the humid tropics of south-east Asia, but this refers only to three runoff plot experiments. Recently Hatch (1981) has reported a study of the protective effect of pepper vines in Sarawak using 10 × 4 m runoff plots under natural rainfall on a 25° slope. One other recent study is that of Van der Linden (1983) in Central Java. Runoff and sediment production were measured on bounded plots 5 × 2 m on slopes ranging from 17 to 20°. The runoff plots

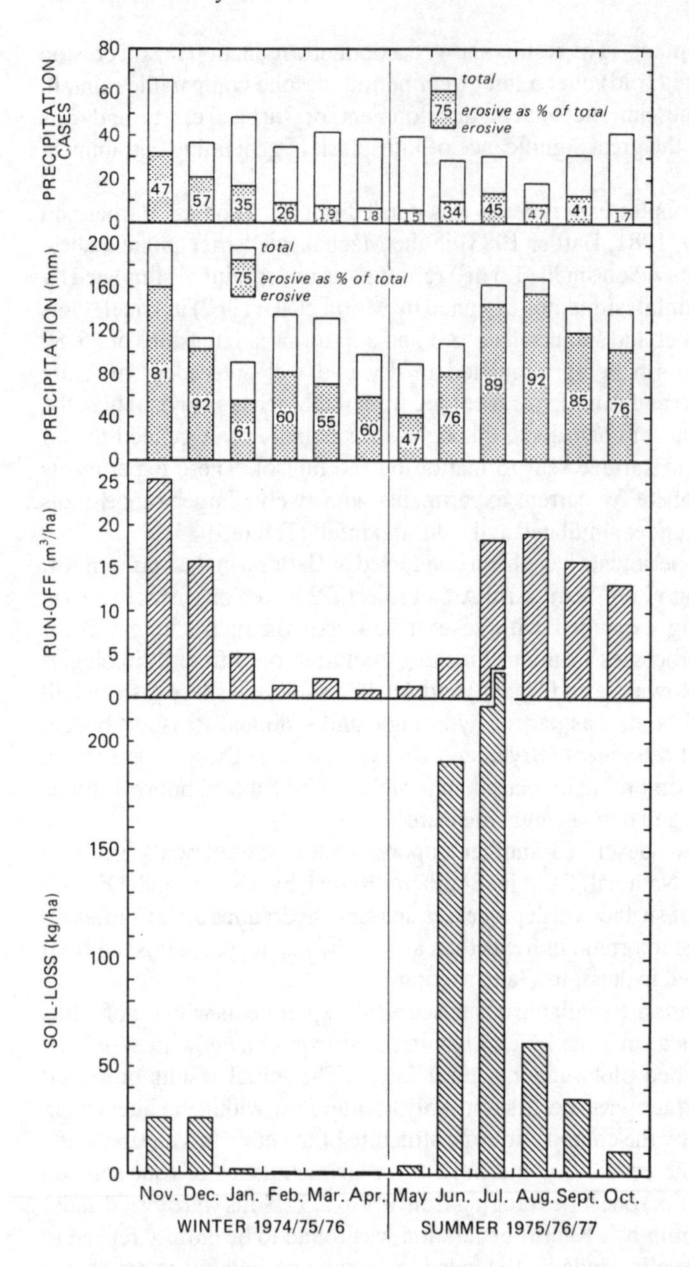

Figure 4.6. Seasonal variation in precipitation, runoff and soil-loss at G. Richter's experimental runoff plots at Trier, Germany (from Richter 1980).

formed an integral part of a study of a 17.8 hectare catchment in which attempts were made to quantify all components of the sediment and water budget, with appropriate adjustment for areas of shifting cultivation and rice paddies. One interesting feature is the remarkably good agreement between areal estimates of annual sediment production based on runoff plot results (13.5 t/ha) and the levels measured at a gauging station (12.4 t/ha) indicating high efficiency of sediment transport in this steeply sloping catchment under intense rainfall averaging 3900 mm/year.

4.2.3 *Experiments in Europe*

In view of the origins of erosion plot research in Europe it is not surprising that there has continued to be a strong emphasis on this approach. Particularly notable is the work of Richter in the Moselle valley, where experiments have been directed particularly to the problems of soil erosion in steeply sloping vineyards. Although much of the work has involved collection of data necessary to apply the USLE methodology, the experiments have been characterized by precise instrumentation and detailed observation of processes, so that they provide rather more information about surface wash processes than most runoff plot experiments. The initial experiments (Richter and Negendank 1977, Richter 1980) were conducted on 13 plots of 2 m width and 8-48 m in length on 20-26° slopes. One plot was under permanent vegetation and the remainder in a vineyard with vines of various ages. Each plot installation included two automatic runoff and sediment traps linked to a central data logger and computer storage, and associated with a meteorological station. Daily observations over a three-year period showed a rather complex seasonal pattern with peak precipitation and runoff in the winter, but peak soil erosion during intense convective summer storms (Fig. 4.6). Observations through the winter season also demonstrated the important interaction between surface wash and soil creep or solifluction induced by melting snow (Richter 1978).

The vineyard experiments described are still in progress, but in 1981 new experiments were started at six field stations on arable land in the vicinity of Trier. These are based on confined plots of 1 m width and 8 m length on standard 8° slopes. Six different soil types were chosen for experiments: regosol on slates, terra fusca on dolomite, palaesol on clay, rendzina on marl, parabraunerde on loess and podzolic braunerde on sandstone. Each was subjected to standardized simulated rainfall of 60 mm h^{-1} intensity from a fall-height of 7-8 m. Details of field plot installations and automatic instrumentation are shown in Figure 4.7. Data from these experiments are still in unpublished form (Richter, 1982) but show relative erosion rates of sandstone > regosol > marl > loess > dolomite > clay.

Messer (1980) reported the results of runoff plot experiments in an Alsatian vineyard. Five plots were used, two on ploughed soil, two on chemically weeded soil and one under permanent grass. The plots were located on a 12° slope and were 3.4 m in width and either 20 or 40 m in length. Observations were carried out under natural rainfall and a strong correlation between peak soil-loss and peak rainfall in July and August was noted, with notably highest soil-loss from the 20 m long chemically weeded plot. Messer (1979) has also reported preliminary results of tests on soil erodibility and runoff generation on small plots under simulated rainfall with 30 min tests at an intensity of 64 mm h^{-1}.

An important series of runoff plot experiments has been carried out by Chisci and his associates based on the experimental research station at Vicarello-Valdera in the Era Valley near Pisa. Runoff plot experiments were started at the station in 1970 with 12 bounded runoff plots of 15 m width and 40 m length on 6° slope. The soils in the area are poorly drained marine clays of low permeability and the initial experiments were designed to test not only different tillage and cropping practices, but also the influence of tile drainage systems. The experiments were extended in 1972 with much larger bounded plots of 15 m and 110 m length on an 11° slope, designed for study of the influence of slope length on soil-loss. All plots were equipped with calibrated weirs, collecting tanks and multislot divisors. The layout of experimental plots is shown in Figure 4.8. The plots were initially used for collection of data under natural rainfall for application of the

Figure 4.7. Experimental installation of G. Richter at Trier, Germany (photos by G. Richter).

USLE system to land use planning. The work included detailed observations of water movement within the soil and provided interesting data on moisture availability to plants from cracking, clay-rich soils and the hydrological importance of soil cracking (Chisci and Zanchi 1981). More recently simulated rainfall studies have also been started with a

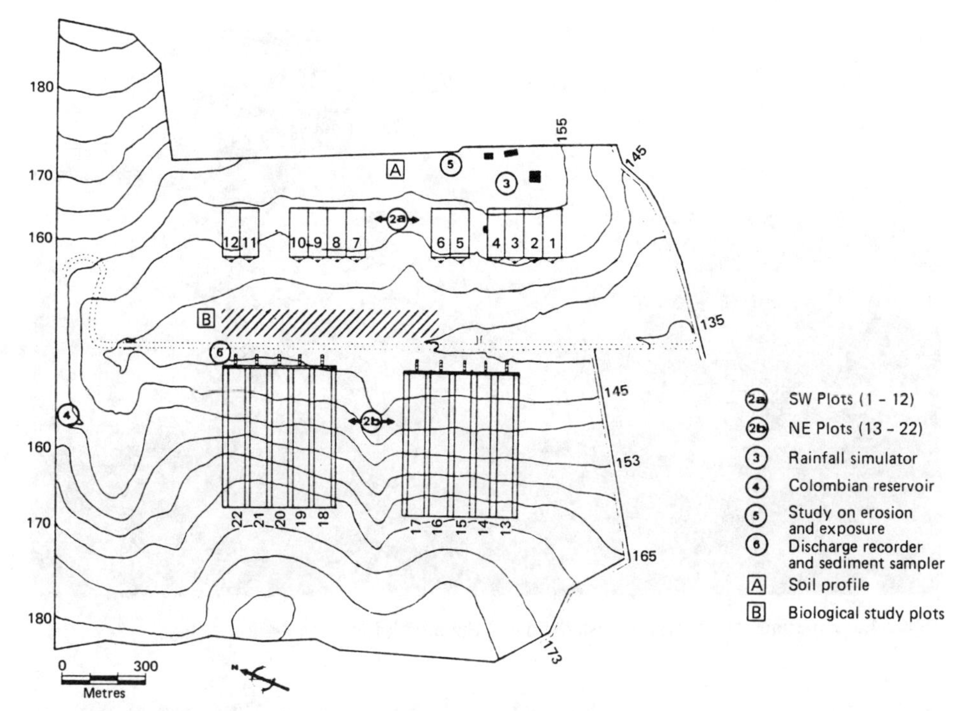

Figure 4.8. Design of experimental runoff plots at San Elizabetta Experimental Centre, Volterra-Pisa, Italy (from Chischi 1981).

large and elaborate simulator capable of delivering rainfall to a total area of 63 m^2 (Zanchi et al. 1981). Experiments have also been extended to instrumentation of small 100 ha catchments in the Era Valley representing a number of typical landscape units. While the runoff plot experiments have now been discontinued, the catchment experiments are still in progress (Chisci and D'Egidio 1981).

The experimental research in the Era Valley has been essentially agricultural in focus, directed towards practical land management problems. Much similar research has been carried out in eastern Europe and one notes particularly the work of the Perieni-Birlad Central Research Station in Romania (Ionita 1983), Djorovic's (1980) work in Yugoslavia, the experiments of Kerenyi, Pinczes and their associates in Hungary (Pinczes et al. 1981, Pinczes 1981) and the observations of Zachar (1982) from many parts of Czechoslovakia. The Perieni-Birlad station was established in 1956 in a severely eroded area, and has been used for experiments to evaluate potential control and reclamation measures. Some of the experiments currently in progress include 13 bounded runoff plots set on a 7° slope on chernozemic soils. Ten of these measure 25 × 4 m or 37.5 × 4 m and are designed for study of the influence of various land use practices on sheetwash. The remaining three plots are 100 × 5 m and are designed for study of rill erosion and incipient gullying. These are equipped with a reservoir and bored pipe which can deliver discharges of 0.6-2 l/sec to the upper edge of the plot. Some of the initial results from the runoff plots show runoff coefficients up to 71% on fallow plots during summer rainstorms, and soil-loss in a single

Figure 4.9. Experimental runoff plot installations at Szymbark Research Station, Poland.

16.8 mm rainfall of 20 t/ha. Other experiments in progress at the station involve study of the hydrologic effect of various reclamation and conservation measures in two small (111 ha and 123 ha) catchments, and studies of gully reclamation.

The studies of Djorovic and his colleagues in Yugoslavia are based in the Srbjan region where erosion is severe. They are based essentially on the methodology introduced by the United States Soil Conservation Service, using runoff plots measuring 20 × 2.5 m. Between 1966 and 1970 more than twohundred runoff plots were established at ten experimental stations through the Srbjan region, covering a wide range of topographic and land-use conditions. Some of the plots are on very steep slopes, notably at the 'Bolec' station where they extend to 30° (Djorovic 1980). These plots have been used to define the following relationships between soil loss, runoff and slope angle:

Grass cover
$$R = 15.4\,x\,0.585 \text{ m}^3 \text{ ha}^{-1} \text{ yr}^{-1}$$
$$E = 0.75\,x\,2.788 \text{ t ha}^{-1} \text{ yr}^{-1}$$

Wheat
$$R = 295\,x\,1.572 \text{ m}^3 \text{ ha}^{-1} \text{ yr}^{-1}$$
$$E = 32\,x\,2.122 \text{ t ha}^{-1} \text{ yr}^{-1}$$

Maize
$$R = 520\,x\,1.183 \text{ m}^3 \text{ ha}^{-1} \text{ yr}^{-1}$$
$$E = 160\,x\,1.163 \text{ t ha}^{-1} \text{ yr}^{-1}$$

where R = runoff, E = soil-loss and x = slope angle (degrees).

Experiments in Hungary have focussed particularly on vineyards in the loess region of Tokaj. Bounded runoff plots measuring 70 × 7.5 m were established and maintained under observation for ten years under natural rainfall. Six different methods of sub-

Figure 4.10. Experimental runoff plot and rainfall simulator of R.-G. Schmidt at Basel, Switzerland (photo by R.-G. Schmidt).

cultivation beneath vines were evaluated for their influence on soil-loss and runoff generation. Over the period of observation a wide range of results was obtained. In general good correlation was measured between soil loss and low intensity rainfall which diminished notably as intensity increased and became non-existent on unprotected plots during the most intense storms (Pinczes 1982). Amongst additional experiments carried out were those at Tokay Nagyhegy using a 200 m plot with average slope greater than 20°. Comparisons between 44 denudation measurement points on the plot and sediment collection at the base showed considerable discrepancies caused by sediment storage on the plot (Kerenyi and Pinczes 1979). This demonstrates the need for closer control and consideration of the slope length factor in runoff plot studies.

Runoff plot research has also been carried out at the Szymbark Research Station in the Eastern Carpathians, Poland by a group from the Polish Academy of Sciences, Krakow, including Froehlich, Slupik, Gil, Welc and Starkel. Several different types of runoff plots were used including standard USSCS plots at Szymbark (Fig. 4.9) and much larger plots of varied shape and area in the adjacent Homerka catchment. The latter were used to determine the influence of slope length on overland flow discharge as part of prolonged detailed studies of all components of water and sediment budgets (Slupik 1981, Froelich 1982). These integrated experiments were maintained in operation for twelve years.

In Switzerland a major project on soil erosion on arable land has been in progress since 1974 which is based in part on experimental runoff plots. Leser (1980) has described the stratified Basel concept of soil erosion research (Table 4.4). The standard runoff plots used are 1-2 m in width and 10-20 m in length. These have been established at four

Table 4.4. Experimental soil erosion research (after Leser 1980).

Dimension	Measurement level/method
Specific stations	1. Test plot
	2. Field trough
Quasi-areal	3. Soil erosion sediment traps
	4. Soil erosion measurement stakes
Areal	5. Complex mapping of erosion damage
	6. Runoff measurement and suspended sediment sampling in a drainage basin

stations (Mohlin, Rothenfluh, Anvil and Langete) in a wide variety of topographic and land-use conditions on slopes of 8-14°. Details of the methodology employed and results from early years of testing under natural rainfall are provided by Schmidt (1979) and Leser et al. (1981). More recently experiments have also been started on 10 × 1 m and 2.5 × 1 m plots using a rainfall simulator with capillary tube drop formers (Fig. 4.10) (Schmidt 1980, 1982).

4.2.4 *Summary remarks on runoff plot experiments*

The experiments described above from various parts of the world provide a representative sample of one type of research related to surface wash being conducted with runoff plots. This research has been directed primarily towards practical land management problems and has focussed on assessment of the relative impact of various land use practices and evaluation of the role of the major environmental factors which affect soil erosion and runoff production. It has been strongly influenced conceptually by the objectives, methods and practical orientation of soil erosion research in the United States, and the USLE approach to erosion assessment. By far the largest amount of research of this sort has been carried out in the United States. It is not possible to examine this work in detail here, but a thoughtful review has been provided by Moldenhauer and Foster (1981). There is no question that this research has been successful in its primary function and it has provided a sound basis for numerous land-use practice recommendations. The studies by Lal and Roose and their associates in Africa, and by Ionita and associates in Romania are good examples of experiments which have led to precise land-use recommendations.

The type of research described was not primarily intended for geomorphic purposes and although many attempts have been made to adapt the results to geomorphology, in general these have not been very successful. The major problem is that the objectives of most experiments have been local and empirical, and they have not been designed to provide the sort of data which can be used successfully in developing geomorphic theory. Two appropriate types of data which can be provided by runoff plot experiments are:

1. Extremely detailed observations and measurements of surface wash and related processes which can be incorporated in physically-based process-response models;
2. Measurements of soil-loss, runoff production and denudation rates under standardized conditions which provide a sound basis for interregional temporal and spatial comparisons.

Most of the runoff plot experiments described do not provide suitable data of the first type either because they are too large to permit very detailed observations during storm events, or because the instrumentation or monitoring system is insufficiently precise to provide detailed information on process controls. The problem can be demonstrated by the requirement for precise rainfall data. In order to model surface wash processes on a runoff plot it is essential to know the relationship between rainfall intensity and infiltration capacity and the temporal and spatial variations over the plot surface during rainstorm events. Very few runoff plot experiments are equipped with instrumentation as sophisticated as that used by Richter at Trier. Many are equipped with one automatic raingauge, but in some cases rainfall amounts are reported only on a daily or less frequent basis. This provides data suitable for gross correlation between soil-loss, runoff and precipitation totals, but not for detailed process analysis. Even where rainfall intensity data are available it is clear that surface wash response varies greatly with soil moisture conditions. This was recognised by Horton in the 1930's when he classified storms into A and B categories on the basis of proximity to preceding rainstorms. Even this basic classification is absent from many runoff plot studies, while many other employ a crude division into 'wet' and 'dry' antecedent moisture conditions. As Bryan (1981) has pointed out this may mask important variations in moisture content which may dramatically affect surface wash and soil entrainment, particularly during critical intense storms of brief duration. If data are to be useful for detailed process analysis more precise characterization of storms is necessary, and preferably frequent or continuous monitoring as in the studies of Richter at Trier and Lal at Ibadan.

Many runoff plot experiments, such as those of Roose in Côte d'Ivoire, have shown that rainsplash plays a critical role in surface wash processes. Complete analysis of rainsplash-surface wash interaction requires data on raindrop size spectra and kinetic energy levels as well as depth of surface water layers and surface microrelief. Raindrop size spectra and energy levels are usually identified in simulated rainfall studies but are almost uniformly ignored in studies under natural rainfall despite the gradual emergence of appropriate instrumentation (Kinnell 1976, Kowal and Kassam 1977, Hudson 1981, Imeson et al. 1981).

Although few runoff plot experiments provide the sort of detailed observations and measurements necessary for analysis of surface wash processes, considerably greater attention has been paid to the need for standardized data collection. The use of standardized 22.1 × 1.85 m plots was fundamental in the US Soil Conservation Service runoff experiments and an important contributor to the value of the data base assembled. Although many of the studies cited above have been designed to provide data comparable to those used in the USLE system very few have been based on plots of standard size. Plots used have varied in length, width and shape, from the 1.5 m^2 plots used by Barber et al. (1979) in Kenya to the 200 × 50 m plots currently used by Lal in Nigeria. Plots have been sited on an extreme range of slope angles, from less than 1° to over 30°. Although local requirements and equipment limitations have dictated the characteristics of plots used, such variation in methodology does introduce problems in data comparison that are seldom explicitly addressed. In most runoff plot experiments runoff production and soil erosion is expressed on a per area per unit time basis. This is based implicitly on an assumption of Hortonian runoff conditions with equal contributions from all parts of the plot, and on uniformly distributed soil entrainment. Such conditions can, of course, exist, but this is probably rather rare. Morgan (1979) has shown, for example, that even on extremely small plots (900 cm^2) most of the soil eroded comes from within 15 cm of the lower plot edge, and that partial area concepts appear to apply even on this microscale.

It is also frequently assumed that the rate of water and sediment delivery is not affected by the length or slope of the runoff plot, and that no storage occurs within the plot. Djorovic (1980), for example, explicitly states that results are not dependent on plot width. In fact long established concepts of catchment hydrology would suggest that the results are influenced by plot shape and this has been shown by Slupik (1979) and by experiments currently in progress in Nigeria (Lal, pers. comm.) and Canada (Luk, pers. comm.). The problem of plot length is more widely recognized and has been addressed by Djorovic (1980) and Moldenhauer and Foster (1981). The plot should be sufficiently long for typical surface wash processes to develop fully. This implies in particular adequate length for rill initiation. Conceptually this is straightforward but in practice it is difficult to suggest an optimal standard length. Moldenhauer and Foster (1981) are probably correct in stating that 1 m is too short, although in exceptional circumstances rills can form in shorter distances. On the other hand, even where surface wash and soil entrainment is active a very long plot may not permit rill development. Dunne (1980) has reported hillslopes of several hundred metres length in Kenya where surface wash is active yet rill incision does not occur. It is important to identify the optimal length, for if the plot is too long, storage of sediment within the plot will almost certainly occur, resulting in a proportionately lower soil-loss per unit area. This means that soil-loss can really only be compared with data from plots of similar length.

Several other problems affect the use of these types of runoff plot data for geomorphic analysis. One is the conversion of a measured weight of soil collected at the end of a plot to a realistic assessment of denudation rate, particularly if the precise derivation of the eroded soil is not known. In practice this can only be approached if the bulk densities of both eroded material and the original soil are known. Only very occasionally are such data reported. Another problem is the general use of plots isolated from upslope influences by a plot boundary. This removes the possibility of sediment replacement from upslope and converts the surface into an erosion surface, although in undisturbed conditions it may have been a transport surface. Clearly unless care is exercised quite erroneous conclusions can be drawn from bounded runoff plot data.

One final obstacle to the comparative use of runoff plot data is the widespread reliance on rainfall simulation. This has greatly increased the flexibility of erosion research and has reduced the time period necessary to acquire useful data. However, unless simulation of natural rainfall is perfect in all respects, direct comparison of results may be difficult. In practice most attention has been paid to intensity, duration and kinetic energy and rather less to drop-size spectra. Frequently the requirements of energy simulation can only be reconciled with reasonable intensity levels by sporadic rainfall application (e.g. in the 'Swanson' boom simulator). Although the simulation of rainfall may appear superficially correct, the interruption in rainfall may significantly affect surface wash processes. Strictly speaking, therefore, runoff plot data collected under simulated rainfall are comparable only with those produced from a similar unit. This question has been discussed in more detail by Bryan (1981). Bryan and De Ploey (1983) have examined some of the difficulties in comparing the results from different rainfall simulators where identical samples and similar testing procedures are used.

4.3 EXPERIMENTAL MICROPLOT AND MICROCATCHMENT STUDIES TO ELUCIDATE SURFACE WASH PROCESSES

In this section attention is directed to experiments which have been established primarily for geomorphic objectives. Little attempt has been made to standardize methodology, but most have included very detailed measurement and observation and collectively have provided a large amount of information on surface wash processes.

4.3.1 *Emmett's contribution*

A number of early studies of surface wash processes were carried out by geomorphologists and hydrologists including Schumm (1956a, b), Young (1958), Schumm and Lusby (1963), and Evenari et al. (1968). Probably the most significant, however, was Emmett's (1970) study of the hydraulics of overland flow on hillslopes. This followed a series of experiments on the hydraulics of thin overland flow carried out by hydraulic engineers such as Izzard (1944, 1946) and Parsons (1949). Another study of overland flow hydraulics was carried out during this period by Iwagaki (1955), but as far as can be ascertained it has been neither translated nor published, and results are not available. Emmett (1970) represented one of the first attempts to incorporate this work and additional experimental data with the general geomorphic concepts introduced by Horton (1945). Emmett's study included detailed laboratory studies using smooth and artificially roughened surfaces under simulated rainfall, which were compared with data from seven natural hillslopes also subjected to simulated rainfall. A flume surface of 4.92 × 1.23 m was used in the laboratory while the field plots were generally 2.15 m wide and more than 12.31 m in length. The same rainfall simulator was used in both tests, providing an average intensity of 198.1-215.9 mm h^{-1} but a median drop size of only 0.5 mm. No attempt was made to measure kinetic energy production, but the simulated rainfall produced clearly diverged widely from natural rainfall of similar intensity. The experiments included detailed observations of flow patterns and numerous measurements of flow depth, velocity and discharge which permitted calculation of all standard hydraulic variables.

Emmett's experiments produced a considerable number of important results only a few of which can be noted here. They demonstrated that most overland flow is within the subcritical regime with Froude numbers typically around 0.2. They confirmed Horton's (1945) view that overland flow consists of both laminar and turbulent flow, and intermediate transitions, but also demonstrated the effect of raindrop impact in retarding flow velocities and increasing resistance, producing 'disturbed' flow with some of the characteristics of both laminar and turbulent flow. The importance of microtopography in controlling flow direction and concentration was noted, and also the effect of organic debris in damming and ponding flow. Active sediment entrainment was observed, but despite clear flow concentration no rill incision was noted. This was particularly noted at the North Fork River site, where erosion was active but no rilling occurred despite 10 hours of continuous simulated rainfall at an intensity of 215.9 mm h^{-1} followed by 6 hours at an intensity of 254 mm h^{-1}. Finally progressive reduction in flow sediment concentration during flow events was noted. This has been explained by other workers (e.g. Lowdermilk and Sundling 1950, Swanson et al. 1965) as reflecting the evolution of an erosional lag pavement, but in Emmett's experiments this did not apply, as an interval in rainfall application was inevitably followed by renewed high sediment concentrations. Emmett suggested that particles were being made available by wetting-drying effects, though the extent of drying in the short interval must have been limited. This result demonstrates the critical influence of small changes in soil moisture content on rates of soil erosion, and emphasizes the inadequacy of conventional reference to 'wet' and 'dry' antecedent moisture conditions discussed above.

The results of Emmett's experiments have been described in some detail because they are central to the geomorphic study of surface wash processes and provided a conceptual framework for a number of experimental studies during the 1970's and for a considerable number of recent and current experiments. Only the latter are discussed here, this section dealing mainly with field experiments while most of the related laboratory research is discussed in the next section. The distinction is rather artificial as much of the research, like Emmett's, has involved both laboratory and field experiments.

4.3.2 *Yair's contribution*

Most of the basic research carried out on surface wash processes during the 1970's involved laboratory experiments. Notable exceptions are the experiments carried out by Yair and his colleagues in Israel which started in the 1960's, and are still in progress. The initial experiments carried out in the Nahal Akhsanya watershed in the northern Jordan

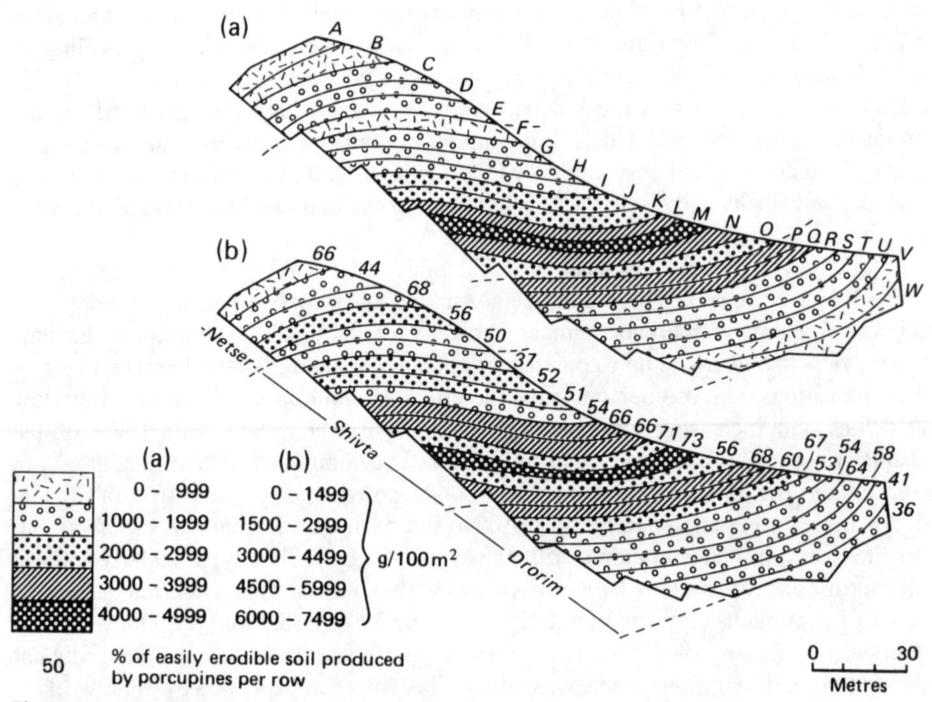

Figure 4.11. Relationship between biological production of erodible soil by porcupines and isopods at Sede Boker experimental site, Israel (from Yair and Lavee 1981).

valley involved instrumentation of a badland talus slope with Gerlach troughs (Yair 1972). Sediment transported by surface wash under natural rainfall was collected for three field seasons, and clearly demonstrated sediment entrainment on the convex divide (Yair 1973). This study was followed by plot experiments on arid talus slopes in the vicinity of Eilat under natural rainfall (Yair and Klein 1973) and later under simulated rainfall (Yair and Lavee 1976, 1977). These studies demonstrated the important influence of heterogeneous surface material properties on surface wash hydrology and talus slope evolution. The most significant experiments, however, were those carried out within an experimental catchment at Sede Boker, northern Negev, which started in 1973 and are still in progress. These have provided detailed data on surface wash processes and runoff response to the sporadic and varied rainfall of the northern Negev. Amongst the more important findings were a systematically higher incidence of rainfall on upper hillslopes which lead to more frequent surface wash and preferential infiltration on midslope sectors, which was reflected in soil moisture contents and enhanced soil development (Yair and Danin 1980). In the Sede Boker experiments surface crusting was found to be a critical control on sediment entrainment by surface wash, and a striking non-uniform pattern of sediment production was noted. Detailed study with a controlled grid network (Fig. 4.11) showed that the dominant factor was the burrowing activity of isopods and porcupines which broke up the resistant surface crust (Yair and Lavee 1981). This result confirms observations of Imeson (1976) in Luxembourg and Roose (1976) in Côte d'Ivoire on the importance of biological activity in erosion by surface wash. In Luxembourg most material transport on investigated slopes was by burrowing animals, while Roose found that material excavation by termites and earthworms greatly exceeded erosion rates on experimental plots. Observations at Sede Boker and in a more extended study throughout southern Israel (Yair and Rutin 1980) showed the location of burrowing to be closely related to soil moisture distribution. It is clear from these studies that biological activity can strongly influence patterns and rates of soil erosion by surface wash. This raises the potential geomorphic significance of seasonal or longer-term fluctuations and cycles in sediment movement. The question of the geomorphic and hydrologic significance of biological activity in relation to surface wash would appear to require considerably more attention. It is notable that this factor has not been considered in any of the large runoff plot experiments described above. It would appear to be of particular importance in agricultural areas where populations and activity are strongly influenced by tillage and manuring practices. Major impact on recorded annual rates of erosion is entirely possible.

The Sede Boker experiments showed that even in an arid environment with infrequent rainfall, erosion by surface wash is primarily weathering- controlled and that variations in surface resistance control sediment entrainment. These may reflect quite complex processes of surface crusting and the impact of disturbing agents on the crust, or, as in the case of Emmett's (1970) experiments, the influence of short term variations in moisture content.

4.3.3 *Belgian contributions*

The importance of time-dependent variations in surface resistance related to surface crusting, particle sorting and the development of erosional lag deposits has been addressed in an important series of experiments in Belgium. The loamy loessial topsoils of Brabant, Flanders and Hesbaye are subject to severe rill and surface wash erosion. In Hesbaye, for example, Bollinne (1977, 1978) reported average rates of erosion of 15 t/ha/yr, an annual denudation rate of about 1 mm/year and peak rates of up to 103 t/ha/yr (6.9 mm/yr) while Gabriels et al. (1977) observed 1.3 mm denudation on an erosion plot in southern Flanders, most of which took place in two months. Although soils are of comparatively homogeneous character, major variations in vulnerability to surface wash do occur, reflecting different structural properties and resistance to surface crusting.

Many experiments have been carried out on the processes of surface crusting and their relationship to soil properties (e.g. Hillel 1960, Bryan 1973). These have shown a general negative correlation between soil aggregate stability or structural properties and crust formation. De Ploey (1977, 1979) found that a $C_{5\text{-}10}$ consistency index, which essentially assesses the volume of added water necessary to cause liquefaction, was an efficient descriptor of the crusting behaviour of loessial soils in Belgium. Surface crusting may influence surface wash processes in two ways, by reducing infiltration capacity and by increasing surface resistance to entrainment. Either may occur independently, although most frequently they are coeval, but the influence on resistance to entrainment appears to be generally of greatest significance on surface wash processes.

De Ploey (1980) and De Ploey and Mucher (1981) have shown that the type and intensity of surface wash processes are closely related to the $C_{5\text{-}10}$ index and to crusting behaviour. On sandy Lier soils, with low $C_{5\text{-}10}$ values, crust development promoted intense wash and repeated micro sand flows precluded the emergence of rills. On the

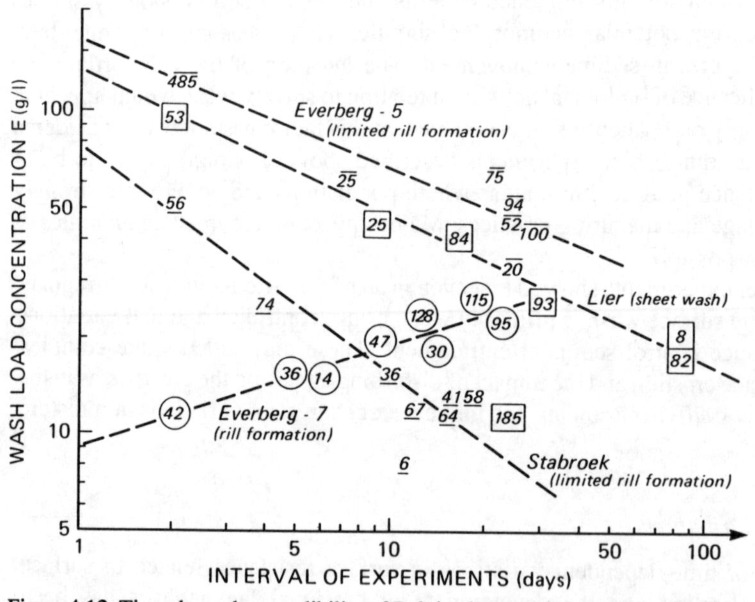

Figure 4.12. Time-dependent erodibility of Belgian loamy topsoils under simulated rainfall at Leuven, Belgium (from De Ploey 1981).

loamy Everberg soils, on the other hand, with even lower $C_{5\text{-}10}$ values, crust occurrence related to clay content and cohesion was critical in the formation and maintenance of individual rills. De Ploey (1980) carried out experiments which demonstrated significant time-dependent variations in the behaviour of different soils subjected to a series of rainstorms separated by intervals of several days. On most of the soils tested, erodibility decreased with time as crusts evolved (Fig. 4.12) but on the Everberg 7 it increased as a result of rill development sustained by crust evolution. Farres (1978) also carried out experiments on crust evolution under simulated rainfall and found that the pattern of change with time was sigmoidal. Initially crust formation was slow as aggregates absorbed water and swelled. Once liquid limit was approached disintegration and crust development was extremely rapid, but eventually with prolonged rainfall it slowed down as the supply of vulnerable aggregates diminished. Clearly the precise pattern of crust development with time will vary greatly between soils, in response to properties such as clay mineralogy, size, shape and porosity of aggregates, organic content and the presence of hydrophobic substances.

The experimental results discussed show the significance of positive or negative time-dependent changes in resistance caused by crusting. These do not, however, explain the short term fluctuations in resistance noted by Emmett (1970). These have also been observed by Bryan (unpublished data) in a laboratory flume study of sheetwash entrainment on cohesive soils. It was found that any interval of more than five minutes in surface wash was followed by a 'flush' of entrainment when flow recommenced. These short-term fluctuations in resistance are believed to reflect very small changes in moisture content, cohesion and surface tension, and are superimposed on the more major positive or negative trends caused by crusting.

In De Ploey's (1980) experiments entrainment took place either as silt and clay particles or microaggregates, but the distribution of particle sizes transported by surface wash changed with distance downslope as micro-aggregates soften, abrade and disperse. A similar effect has been noted by Imeson and Jungerius (1976) in a field study of colluvial processes in the Luxembourg Ardennes. Luk (1983) also noted time-dependent changes in the size of aggregates entrained by sheetwash and rainsplash on laboratory microplots under simulated rainfall. It should be noted, however, that the trends and fluctuations in surface resistance noted above reflect changes in soil structure rather than selective particle entrainment. Selective particle entrainment and the evolution of a resistant lag erosional pavement (as opposed to a surface crust) has been observed in other experiments. Savat and Poesen (1977) showed the effects of selective transport by splash and discontinuous runoff in experiments which related tests on Kalahari sands in a laboratory at Leuven to field processes near Kinshasa in Zaire. Discontinuous runoff transported preferentially all coarse grains larger than 0.22 mm diameter, while rainsplash transported only material between 0.125 and 0.6 mm. This resulted in a lag deposit of coarse grains on the hillcrest where only splash is active, while on steeper slopes discontinuous runoff resulted in an erosional pavement of finer grain size than the original soil.

These results were expanded in a later laboratory study under simulated rainfall which included sandy and loessial samples from Belgium (Poesen and Savat 1980). When rainsplash occurred alone the grains preferentially eroded typically reflected modal size classes, coarser particles remaining on the surface, and finer particles being washed down to form a filtration pavement at 2-5 mm depth. When discontinuous runoff was present, coarser grains were transported more easily than average grain sizes and clear sorting occurred. Even more pronounced sorting occurred when continuous laminar afterflow was used. It is notable that in both cases flow was laminar and supercritical with Froude numbers ranging from 1 to 3.43 depending on slope angle, well above the values observed in Emmett's (1970) experiments.

De Ploey and his associates at Leuven have been involved in a considerable number of important experiments on surface wash processes but as these were primarily based on laboratory research they are discussed in the next section.

4.3.4 *Canadian-Israeli cooperation*

The field experimental approach adopted by Emmett was used in a number of experimental studies in the late 1970's, all more strictly geomorphological in focus than most of the runoff plot studies described. In the Dinosaur Badlands of western Canada Bryan et al. (1978) carried out simulated rainfall tests on microcatchments of diverse lithology, in which test areas were demarcated by natural drainage lines. While most surfaces eventually generated Horton overland flow, major differences in runoff thresholds, runoff coefficients and surface wash patterns between different lithologies were noted, as it was demonstrated that hydrological concepts of partial area contributions apply on shale surfaces even at this microscale. Rillwash was also shown to be controlled by variations in surface material properties and to be closely related to flow in shallow intermittent surface pipes.

The initial work in the Dinosaur Badlands was followed by more detailed experiments

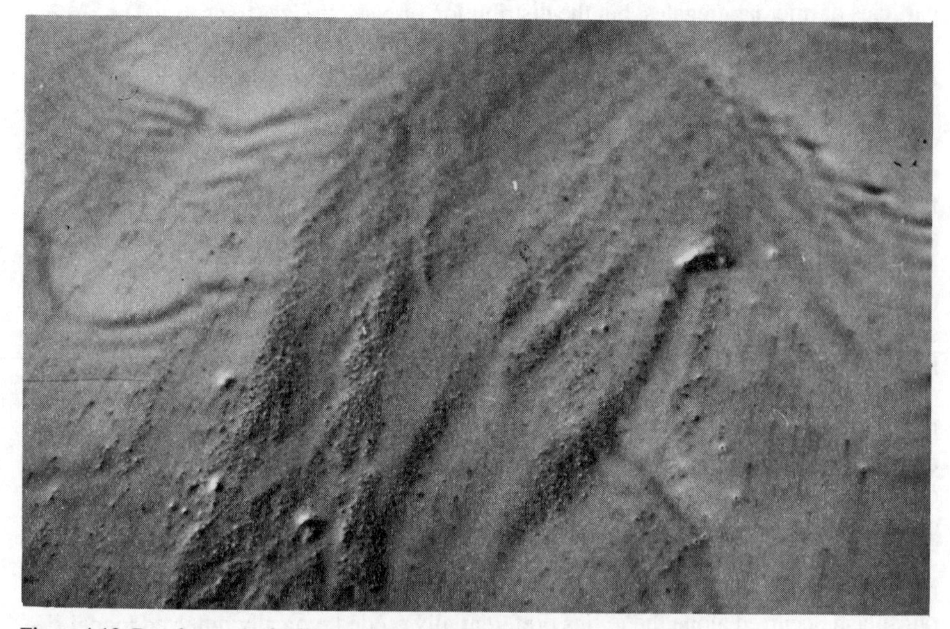

Figure 4.13. Development of roll waves and scoured chutes in surface wash on miniature pediments in Dinosaur Badlands, Alberta, Canada (photo by W. K. Hodges).

on a wider range of lithologic units in which material response to wetting and its influence on surface wash processes were precisely characterized (Hodges and Bryan 1982, Hodges 1984, Bryan and Hodges 1984). These experiments were of particular interest as one of the early attempts to link detailed study of surface wash hydraulics and sediment entrainment patterns to the magnitude and frequency of surface wash events through the relationship to long-term climatic records. It also provided some of the first detailed observations of surface wash processes controlled and complicated by sodium-dominated swelling montmorillonites. The area was shown to be one of very high (though varied) erosion rates with average denudation rates by surface wash of about 4 mm per year. This rate agrees extremely closely with average figures derived independently from 'bedstead' studies (Campbell 1982) and instrumented catchments (Bryan and Campbell 1986).

The most detailed hydraulic studies were carried out as part of a study of the relationship between surface wash processes and the origins of micropediments (Hodges 1982). These showed the same character and localization of surface wash on shale and sandstone surfaces noted in earlier experiments, but on the gently sloping micropediment a complex sequence of flow patterns was observed. Initially flow covered the complete pediment surface with Reynolds numbers in the laminar to transitional range. As the hydrograph peak approached erosional chutes developed associated with roll waves and turbulent flow. As flow receded hydraulic instability features disappeared, flow reverted to laminar (transitional) and chutes were eliminated by sedimentation. Only on shale surfaces was flow clearly subcritical with Froude numbers as low as those in Emmett's (1970) study. On other surfaces even in sheetflow some values moved into the supercritical range, particularly in tests carried out in 'wet' antecedent moisture conditions. Rill flow on all surfaces was typically supercritical with Froude numbers ranging from 0.98 to 5.56. The development of erosional chutes on the pediment coincided with roll-wave development at Froude numbers around 2.5 (Fig. 4.13). Manning n values showed differences up to fourfold between values on different surfaces, but the roughest surface (the shale) showed notable hydraulic smoothing between 'dry' and 'wet' antecedent moisture tests.

A comparative experimental study using a similar microcatchment approach was carried out by Yair et al. (1980b) in the Zin Valley badlands of the northern Negev desert, Israel. Although the superficial appearance was similar to the Dinosaur Badlands, the parent Taquiya shales were of more homogeneous lithology. The tests carried out included the equivalent of a 200-year storm yet the erosion rates observed were miniscule compared with those in the Canadian badlands. This was attributed to the mixed clay mineralogy of kaolinite and montmorillonite and strong flocculation by calcium carbonate and gypsum. Surface wash occurred but pipeflow and sporadic mudflows obstructed detailed hydraulic measurement. Despite the homogeneity of parent materials and the low erosion rates marked differences were noted between response on north and south-facing slopes reflecting the influence of microclimate on surface crust evolution and lichen colonization.

4.3.5 *Anglo-American contributions*

Dunne and Dietrich (1980a, b) used rather similar methodology in an experimental study of infiltration, runoff generation and surface wash hydraulics on gentle slopes in southern

Kenya. Rainfall was simulated by downward pointing nozzles which produced rather large drops (2-2.7 mm diameter) though the median drop size and kinetic energy reproduction compared well with other units (Dunne et al. 1980). Confined plots of 5 × 1-2 m were used rather than natural microcatchments. The slopes selected covered three different soil types and various different vegetation conditions, but all were free of visible rill incision, implying absence of flow depth variations. The rainstorms simulated varied in natural recurrence interval between 4 and 200 years. Very different response was noted on the three soils, but in general it appears that natural runoff production is rather infrequent. This is due in part to the significance of progressive surface crusting on infiltration, and, on swelling vertisols, the influence of intense desiccation cracking on infiltration. In sufficiently prolonged rainfall these cracks close, and infiltration drops to a very low level.

On vegetated vertisol plots, vegetation and micro-topographic variations concentrated flow into narrow zones (cf. Emmett 1970). This was true to a lesser extent of the luvisolic plots in Amboseli, while the unvegetated plots on Kilimanjaro lava produced continuous surface wash, though some zones of concentration and greater depth were noted. The surface wash observed was laminar or transitional and disturbed with Reynolds numbers in the same range as those in Emmett's (1970) experiments, though some much lower values were noted, particularly on the vegetated vertisolic plots on which higher Darcy-Weisbach friction factors were calculated. No fully developed turbulent flow was observed. Hydraulic parameters were used to compute hillslope hydrographs with the kinematic approximation method adapted for recessional infiltration. These demonstrate important variations in hydrograph shape and peak discharge related to differences in gradient and surface resistance. The results of the study show that under natural rainstorm conditions most of the runoff generated on hillslopes does not contribute to channel flow but does contribute to variations in soil moisture recharge and plant production. This confirms the empirical evidence of Yair's experiments at Sede Boker, though evidence of a relationship to systematic variation in pedogenic development in Kenya has not been reported.

Morgan (1980) adopted a different approach in an experimental study of sediment transport by surface wash on sandy soils in Bedfordshire, England. This study was carried out with Gerlach troughs placed en echelon down a slope profile, so that effective slope length ranged from 172 to 201 m. The sandy soils were in fallow, periodically worked by rotovation so that the vegetation cover varied from 0 to 25%. Earlier studies (e.g. Morgan 1978) had shown the soil to be highly erodible producing high sediment concentration in surface wash (up to 70 g l^{-1} at peak flow) and possibly liquefaction and micro mass-wasting. Natural rainfall was used and sediment and discharge totals on an individual storm basis, associated with detailed observations of flow patterns during storm events. The object of the study was to attempt to link flow hydraulic characteristics with sediment transport and to test the applicability of a number of sediment transport formulae developed originally for open channel flow. This was attempted by back-calculating hydraulic parameters from discharge totals and observations of the location and duration of flow rather than direct instantaneous measurement. Although this approach is obviously less accurate, it is potentially very useful as such total data are more frequently available than instantaneous measurements. The storms analysed were all of brief duration and produced discontinuous surface wash in anastomosing rivulets like those noted by Emmett (1970) in Wyoming. Calculated Reynolds numbers ranged from 0.34 to 74, well within the laminar range, though all the flow was affected by raindrop impact and surface roughness and was classified as disturbed flow. The calculated Froude numbers (0.003 to 0.03) indicated sub-critical conditions, but as these are averaged rather than instantaneous values it is quite possible that significantly higher values occurred, especially in rivulet flow. The sediment transport formula eventually derived was based on Engelund's sediment transport capacity equation (Engelund 1967) and a modified form of the Manning flow velocity equation. Sediment yield predictions using the formula were significantly correlated with observed results but it was suggested that predictive capacity could be significantly increased by improved characterization of surface roughness conditions.

4.3.6 *The replication problem and modelling*

While some of the most important recent field studies of surface wash processes have concentrated on characterization of hydraulic conditions, very little attention has been paid to the reproducibility of surface wash erosion results and the need to replicate experiments. Bryan and Luk (1981) carried out replicate laboratory tests on a small (30 × 30 cm) plot under simulated rainfall. Twenty replicate tests were completed for each of three soils, all major contributory factors (e.g. rainfall intensity, duration, slope angle, soil type, cover type, initial moisture content, bulk density, etc.) being kept constant. Despite this coefficients of variation ranging from 11 to 25% were measured for various soil loss parameters. Measurements were made at 5 min intervals through the tests, and variability was noted to decrease significantly with storm duration. This emphasizes the desirability not only of replicating tests, but also increasing test duration. Comparative analysis of various soil properties and test parameters indicated that most variability could be attributed to surface roughness and to aggregate stability. This study was followed by a series of field experiments using similar plots under simulated rainfall on a variety of soil types, but using soils in-situ in the field with natural soil structure (Luk and Morgan 1981, Luk 1982). As this introduced elements of spatial soil variability as well, much higher coefficients of variation were measured, ranging from 21-35% for rainwash, 20-75% for runoff and 13-93% for sediment concentration. Significant differences were also noted between the results of two operators despite identical test procedures. These two studies raise points of considerable practical importance for surface wash research, and particularly emphasize the difficulty of comparing directly results obtained with different operators, rainfall simulators and test procedures.

The second stage of the project (Hamilton 1984) involved simulated rainfall tests on 1.8 × 4.3 m^2 plots at two sites in southern Ontario, Canada. Both soils tested were agriculturally disturbed loams, but with markedly different aggregation characteristics. The eleven replicate tests respectively were carried out, and as in the earlier tests, variability was found to diminish with storm duration. Nevertheless even at the end of sixty minutes of rainfall variability remained high, ranging from 68.5-105.9% for rainwash and 79.8-89.3% for runoff.

In an attempt to isolate factors contributing to variability detailed attention was paid to the characterization of antecedent soil moisture conditions, measured both gravimetrically and by neutron probe. Although spatially variable, coefficients of variation were much lower than those for rainwash and runoff. Antecedent moisture levels were significantly related to soil loss by rainwash, but showed perceptibly weaker correlation with runoff. These results support Emmett's (1970) conclusion on the critical role of changing soil resistance in relation to soil moisture content. Attempts were made to monitor changes in soil moisture content during test storms with a soil psychrometer to see if a more explicit relationship could be established between moisture content and soil entrainment. In some tests a marked increase occurred in runoff and soil loss which could be related to the point at which the soil reached saturation. In most tests, however, any clear relationship was obscured by the time lag between entrainment and the arrival of sediment at the end of the plot, irregular storage of both sediment and water on the plot, and by the erratic performance of the psychrometer, particularly at extreme moisture levels. It does appear from those experiments that the interaction of soil moisture content and soil erodibility is critical in surface wash processes, and any characterization of soil erodibility must encompass some reference to change with moisture content. These would appear to be a clear requirement for further, more precise experimentation if these interactions are to be fully identified.

Scoging (1982) has also approached the problem of variability in surface wash in sprinkling experiments carried out in the semi-arid region of south-eastern Spain. Most of the rainfall in the region is of comparatively low intensity and runoff occurs not as Horton flow, but as a result of the very limited storage capacity of shallow soils (Scoging and Thornes 1980). The experiments involved repeated sprinkling on 1 m^2 runoff plots set in a framework of larger plots (1.48 × 9.88 m), on which detailed observations of erosion and deposition under natural rainfall were collected over a twelve month period using a grid of nails and washers. These measurements showed extreme variability with patches of denudation of up to 27 mm depth, and deposition to 12 mm depth. Complex runoff response and erosional patterns with significant differences between coarse and fine sites were noted. Runoff response was more closely related to the time-decay function of the infiltration capacity curve than to the final infiltration capacity. Higher runoff coefficients occurred at fine sites, but, except in low intensity rainfall, the coarse sites produced higher sediment concentrations. This was attributed to lack of cohesion between particles, and the evolution of a resistant surface crust on fine sites during higher intensity rains. It is notable also that runoff coefficients on all sites decreased slightly with rainfall intensity. Scoging (1980) also used the results of these experiments to test a soil erosion model incorporating equations for infiltration with time decay, lateral inflow, continuity of flow over time and space, and a flow equation.

Bork and Rohdenburg (1979, 1980) also carried out extensive sprinkling tests on 1 m^2 plots in Almeria Province, south-eastern Spain. Very much higher rainfall intensities were used with 147.7 mm h^{-1} compared with 44 mm h^{-1} in Scoging's experiments. The experiments included detailed measurements of overland flow, soil loss and soil moisture suction at 5 and 10 cm depth. A wide variety of hydrographs was observed on the 100 test plots:

a) Overland flow reaches constant level within 60 minutes after continuous increase;
b) Continuous increase throughout 60 minutes without reaching constant level;
c) Steady rise to early peak then decrease to constant level;
d) Very irregular pattern.

Stepwise regression was used to analyse the influence of thirteen potential contributory factors. The most significant controls were found to be the precipitation rate and the macro pore volume at 5-10 cm and 40-50 cm depth.

Scoging's study in Spain represents a rare attempt to link field experiments on surface

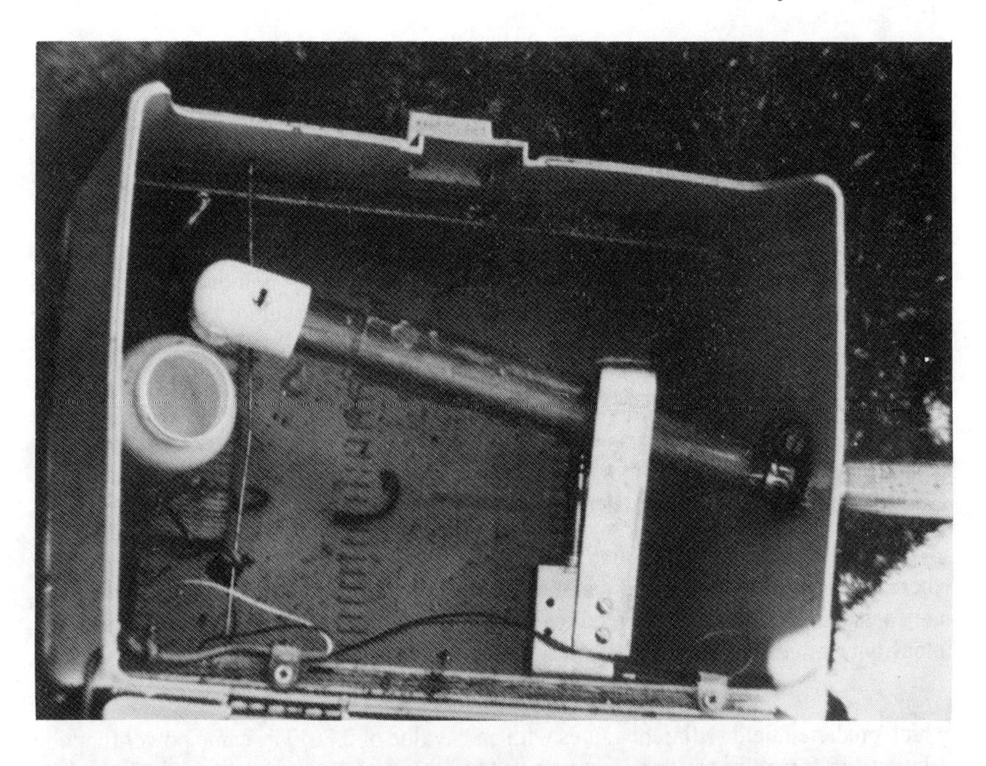

Figure 4.14. Automated sampler in use at experimental site of L. Band, California, USA (photos by L. Band).

wash directly to theoretical and mathematical models. Band (1983, 1985) also used this approach in a study of slope development in northern California. The site, a hydraulic gold mine abandoned more than a hundred years ago, was chosen because initial boundary conditions could be established fairly confidently, the slopes developing on weathered phyllite are simple and unrilled, erosion is rapid and measurable in a short time, and a detailed rainfall record back to 1871 is available. The experiments were set up on short (< 6 m), steep (30-35°) rectilinear slopes without vegetation to provide data for testing a model for sediment transport by surface wash, and for computer simulation of slope development. Water and sediment discharge were measured with an automatic sampler (Fig. 4.14) which simultaneously collects samples for chart periods (about 5 min) at various locations at intervals during storm events. Initial boundary conditions were established by historic photographs, dating of trees growing on or near drainage lines, and morphologic evidence. The rainfall record is sufficiently detailed to allow separation into discrete intensity/time units as a basis for magnitude-frequency analysis.

The sediment transport equation used was given by Carson and Kirkby (1972) as:

$$q_s = K_h \, q_w^m \tan^n \theta$$

where q_s and q_w = sediment and water discharge, θ = slope angle in degrees, and K_h = a rate constant. The data set collected over two months under natural rainstorms of varied intensity resulted in the following estimate of parameters:

$$q_s = 0.000025 \, q_w^{2.07} \tan^{0.84} \theta$$

which predicted field sediment values with an r^2 value of 0.80. The same power function relationship for sediment transport was used with a simple rainfall-runoff relationship and a descriptive equation for soil creep to derive a dimensionless numerical simulation model. With appropriate values for erosion by surface wash and soil creep derived from field experiments this generated a profile which closely matched the lower part of the field slopes, but did not predict the crestal convexity accurately. When the rate constant for soil creep is doubled, to reflect the effect of rainsplash and raincreep, not included in the model, crestal prediction is much improved. The model produced is sufficiently sensitive to permit identification of the individual contribution of discrete rainfall intensities to total soil loss and hillslope form development. This allows precise determination of the 'maximum work' event in magnitude-frequency analysis, which was found to accord well with Pearce's (1976) results for surface wash, and those of Wolman and Miller (1960) for river systems.

4.3.7 *Summary remarks*

While the preceding discussion of microplot and microcatchment is certainly not exhaustive, it does provide a reasonable cross-section of recent experimental field research on surface wash processes employing this methodology. The main emphasis has been on detailed observations of infiltration patterns and hillslope hydrology and on analysis of the hydraulic characteristics of surface wash and the applicability of hydraulic relationships derived from fluvial geomorphology. Increasingly it has been recognized that surface wash processes and their contribution to landform evolution are strongly affected or controlled by soil erodibility and the behaviour of the soil surface. Important aspects include:

1. Variability of initial textural and aggregation character;

2. Spatial and temporal change caused by selective erosion leading to the dynamic textural catena noted by Savat and Poesen (1977) and Scoging amongst others;
3. Temporal changes in erodibility related to changes in soil moisture content;
4. Time-dependent evolution of surface crusts in response to rainsplash activity and other processes;
5. The role of soil fauna in disturbing surface crusts and providing sediment vulnerable to entrainment by surface wash.

Most of the experiments carried out on the dynamic behaviour of the soil surface and its relationship to surface wash processes have focussed on the physical behaviour of the soil system and have viewed the entrainment process largely as a physical or hydraulic process. Increasingly however, geomorphologists have recognized that many of the soil properties which affect surface wash processes are a result of chemical as well as physical processes. The dynamics of soil aggregation and surface crusting, and the activity of soil fauna, in particular, are closely linked to soil organic matter, to biochemical processes and to nutrient cycling. It has also been recognized that the processes of entrainment and transportation in surface wash also reflect a balance between soil chemistry and that of surface and subsurface water. As a result, in recent years, a number of experimental studies have focussed on characterization of soil chemical characteristics and solution, as well as physical processes.

Amongst early experiments on the role of chemical processes in surface wash were microplot and microcatchment studies carried out by Ponce (1975) and Ponce and Hawkins (1978) on Mancos Shale in Utah. The Mancos Shale, which outcrops in the south-western United States, is a complex montmorillonitic unit with gypsum and calcite veinlets, thin shaley limestones and sandstones. The more shaley members form areas of typical badland geomorphology. The Mancos Shale has a variable, though typically high, soluble mineral content and contributes much salinity and sediment to drainage systems in the area. The studies in Utah were carried out with several different rainfall simulators on a variety of small plots from 0.23-3.7 m^2 in area. Distilled water was used in simulated high-intensity rainstorms of 60 min duration during which runoff discharge, sediment concentration and solute concentrations were measured at five minute intervals. The significance of using distilled water on soils with high soluble mineral content has been demonstrated by Imeson et al. (1982) in applying Kamphorst and Bolt's (1976) graph to the development of piping and gullying in the Beni Boufrah badlands of Morocco. This graph relates the balance between exchangeable sodium percentage in soil to the electrical conductivity of the soil solution, and shows the effect of this on dispersion, flocculation and swelling of the soil. It is clear from this graph that use of distilled water on regolith with high soluble mineral content could be expected to produce maximum soil dispersion and abundant release of sediment for entrainment by surface wash. In the Utah experiments, sediment concentration patterns were fairly well-defined, with a general decrease with storm duration. This could be attributed to rapid dispersion early in the rainstorm with the first influx of water of low specific electrical conductivity to the soil, followed by rapid evacuation of released fines. It is possible, however, that physical surface crust evolution was also involved and the effects are not easy to separate. Solute concentration patterns were more complex, in some cases showing decline, in some cases increase, and frequently irregular fluctuations. One interesting result is that both solute and sediment concentration were found to increase with plot size. This was attributed to the greater erosive capacity of rills developed on the larger microcatchments, which resulted in higher particle entrainment and exposure of subsurface regolith where the soluble mineral content is more than twice as high.

The increase in solute concentration with plot size can reflect higher entrainment and more saline regolith, but it may also reflect time-dependent dissolution of sediments in transport. Laronne (1981) has carried out a detailed study of the dissolution kinetics of alluvium derived from Mancos Shale, and has related this to field experiments on Mancos Shale hillslopes in Colorado (Laronne 1982, Laronne and Shen 1982). In these experiments water of higher electrical conductivity was used (450 $\mu S\ cm^{-1}$ at 25°C) and flow was generated by a perforated pipe on the surface. Sites were chosen with well-developed simple rill systems on slopes ranging from 5-28°, of length 12.2-76 m, with vegetation cover ranging from 1-10%. Samples were collected from the leading edge of flow and at increasing intervals up to a total duration of 60 minutes. They were also collected at different distances downslope. One would expect the differing chemical balance and absence of rainsplash to produce lower sediment concentrations than in the Utah study, and indeed this was the case for samples near the slope crest. Those influences were counteracted by the greater length and steepness of slopes, and the effectiveness of rill erosion, so that sediment concentrations increase downslope. As in the other tests, there was marked variation between sites, some showing higher initial conductivities and sediment concentration followed by steady decrease to a low level by the end of the test. Others showed irregular increase in both with several peaks. There was a clear but not constant relationship between conductivity and sediment concentration. In some cases peak conductivity preceded peak sediment concentration, while sometimes the reverse was true. In general all sites showed increased conductivity of water downslope. This is partly a reflection of increased sediment concentration and of rill incision through to the subsurface, but also of greater dissolution of sediment transported in surface wash.

The Utah and Colorado experiments demonstrate the considerable complexity of the relationship between solute release, physical entrainment, and solute and sediment concentrations in surface wash, even on hillslopes developed on more or less homogeneous regolith. A more complex situation was addressed in experiments carried out in the Dinosaur Badlands of Alberta, Canada, by Bryan et al. (1984). This environment is also dominated by sodium-rich montmorillonites, but most hillslopes of any size are composed of a series of different shale beds, some sandstones, resistant siderite bands and occasional micropediments. Considerable differences exist in the character and intensity of rill system development (Bowyer-Bower 1984, Bowyer-Bower and Bryan 1984), and the situation is further complicated by a very close interrelationship between sheetwash, rillwash and shallow pipe flow.

The methodology for those experiments was the same as for runoff experiments described earlier with rainfall simulation on microcatchments of varied lithology with areas of 16-35 m^2. Samples were collected at irregular but frequent intervals, sometimes reaching 2-3 per minute, specific electrical conductivity being measured immediately in the field. As in other experiments the pattern of conductivity and sediment concentration was very varied, but most sites showing a flushing effect as salts precipitated on the surface between storm events were removed in the first minutes of flow. It is also notable that sediment concentrations, while variable, were generally much higher than in the Utah and Colorado experiments, reaching peaks of 300 $g\ l^{-1}$. Marked differences in conductivity and sediment concentration were noted between sites and between rills draining different lithologic units within the same site. With the assistance of laboratory studies of regolith dissolution kinetics it was possible to identify lithologic units in terms of total amount and release rate of solutes. This presents the possibility in future studies of identifying flow source areas by conductivity levels. The situation is complicated, however, by marked temporal and spatial three-dimensional variations in electrical conductivity within the same lithologic unit (Sutherland 1983). A further complication is the involvement of both shallow and deep pipe and tunnel flow. Recent observations (Bryan and Harvey 1985) have shown that such flow typically has higher solute concentrations than surface flow and contributes disproportionately to the solute load of ephemeral drainage systems.

The three sets of experiments described were all preliminary in nature and all were carried out in areas dominated by dispersing, swelling clays with high soluble mineral concentrations. The interaction of physical and chemical processes, and their influence on sediment concentrations in surface wash are probably extreme in these conditions. Nevertheless the studies do indicate the potential importance of the interrelationship between soil chemistry, water chemistry and the characteristics of surface wash. These should not be ignored in future experiments on surface wash, even in less extreme environments, and should be explicitly examined over a wider range of pedologic, geomorphic and climatic conditions.

4.4 LABORATORY

4.4.1 *Early work*

The tradition of laboratory research on surface wash is almost as old as that of field experimentation and, as noted in the introduction, both approaches have been closely linked in the evolution of concepts about surface wash processes and their geomorphic and hydrologic significance. A number of significant experiments were carried out during the 1930's and 1940's, primarily by agricultural engineers and soil conservationists in the United States. More recently some of the most significant research has been carried out by hydraulic engineers and this is reviewed by Emmett (1970) in relation to his own laboratory experiments on surface wash hydraulics, discussed above. It was not until the 1970's that geomorphologists started to play a major role in laboratory research on surface wash processes. Particularly important during this period has been the work carried out by De Ploey and his associates at Leuven, and Schumm and his associates at Colorado State University, Fort Collins.

4.4.2 *Laboratory experiments at Leuven*

Over the past fifteen years extensive experiments on many aspects of surface wash processes and their relationship to surface material properties have been carried out in the Laboratory for Experimental Geomorphology, Leuven. This work developed from initial field studies of slopewash in Zaire (De Ploey and Savat 1968, De Ploey 1969) and has more recently been related to field research in Tunisia, Nigeria and Brazil, as well as Belgium. Most of the work has used small laboratory flumes ranging from 1-4 m in length under rainfall simulators with capillary tube drop formers. Earlier experiments were limited by an available fall-height of 2.1-2.5 m but since the late 1970's the work has been carried out in a laboratory with available fall-height of 7.1 m, allowing excellent reproduction of the terminal velocity and kinetic energy of natural rainfall. The experiments have been characterized by detailed observations and measurements of the behaviour of surface materials and the hydraulic characteristics of flow, and have included a considerable number of innovations in methodology. In a study of the micromovements or the influence of runoff, for example, laser beam reflection was used to demonstrate upslope creep resulting from vortex erosion (De Ploey and Moeyersons 1975, De Ploey et al. 1976). In another experiment Savat (1976) used thin silver discs for precise photographic measurement of flow velocities. This was the first of an important set of experiments on surface wash hydraulics continued until Savat's untimely death in 1982. The experiment was designed to test the applicability of standard theoretical flow formulae to the conditions of surface wash over a fine loess surface on slopes ranging from 0.5 to 8°. Generally satisfactory results were obtained with the Hagen-Poiseuille law for laminar flow and the Manning and Chezy formulae for turbulent flow. On gentle slopes flow was found to be laminar but on 8° slopes turbulence was fully developed. Various suspended sediment load and total load formulae developed for river flow significantly under-predicted actual wash loads, which typically increased as a logarithmic function of slope, rather than a power function as in most river formulae. Savat also described the development of scour-steps in the middle of the flume (after 2 m flow length) as the first stage in rill generation. These developed once wash-loads exceeded 35 g l^{-1} and were associated with the appearance of standing waves and the initiation of undulations in the loess bed. Adjacent scour steps typically developed at intervals of 0.2 m.

One major difficulty encountered in this initial study was accurate measurement of flow depth after instability occurred and surface waves formed. This prevented accurate

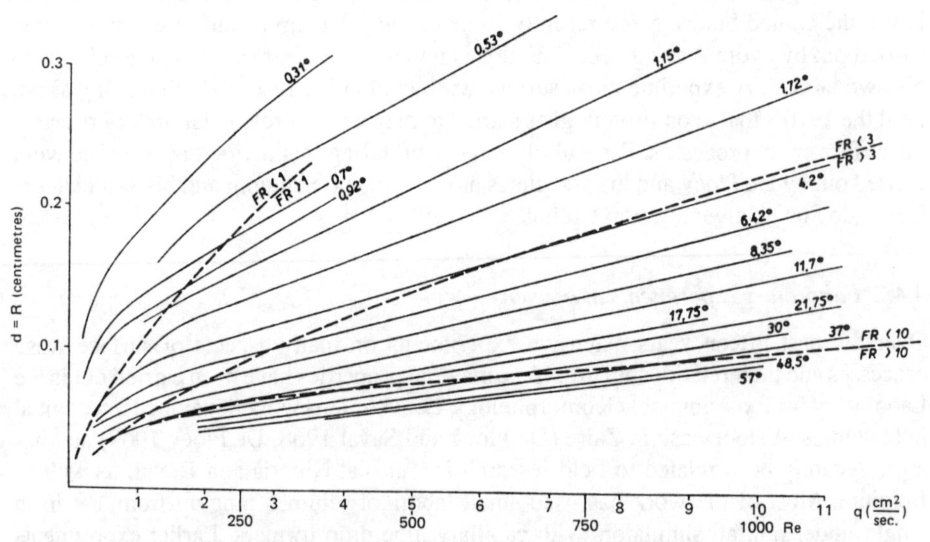

Figure 4.15. Depths and Froude numbers as a function of slope and unit discharge or Reynolds number in laboratory hydraulic experiments at Leuven, Belgium (from Savat 1977).

calculation of standard hydraulic parameters such as friction factors, Reynolds and Froude numbers and shear stress. In a subsequent study without a loess bed (Savat 1977) this problem was approached by using the weight of flow to calculate depth, after appropriate correction for the acceleration of flow to steady velocities. With the small flume used in this experiment it was possible to determine the relationship between flow depth, unit discharge and slope angle for slopes from >1° to 57° using flow generated by a header pipe, with and without simulated rainfall. This enabled calculation of hydraulic parameters and characterization of the condition of transition from laminar to turbulent flow as shown in the basic data set (Fig. 4.15). Virtually all flows were in the supercritical regime, except on the most gentle slopes. Laminar, transitional and turbulent flow occurred on all slopes but the transition appeared to occur more swiftly on steeper slopes. It also varied with flow depth, the breakpoints between laminar/transitional/turbulent being Re 300 and 685 respectively for 0.1 cm deep flows and 340 and 790 for 0.2 cm flows.

The study of the impact of rainfall on flow hydraulics was limited as only one intensity (60 mm h^{-1}) was used and the limited fall-height then employed prevented accurate reproduction of terminal velocities. The results agree with earlier studies in showing diminished flow velocity and increased depth and friction factors. Earlier studies were carried out on very low slope angles. Savat's results show that on gentle slopes increase in flow depth is only 6% and in friction factor only 20%. Both effects diminish with increased slope angle, and on slopes above 45° both flow depth and friction factor were actually decreased by rain drop impact.

The same basic methodology, but without rainfall, was employed in another experiment on friction factors in shallow flows on steep slopes (Savat 1980). Data from a reference bare plastic flume were compared with those from six other flumes with roughened surfaces of varied dimensions produced by varnished sand and loess. These experiments showed that the ratio of surface velocity to mean velocity increased with roughness and slope at Reynolds numbers below 2440. This was interpreted as indicating a laminar regime superimposed on a turbulent sub-layer. Standard hydraulic formulae were found to underestimate friction factors.

Savat's experiments on the hydraulic characterization of sheetflow are well-known and their importance for the development of surface wash theory is widely appreciated. Less well-known are some of the concurrent experiments carried out on selective particle transport, rainsplash processes and particularly those on rill initiation. Savat's initial experiments on rill initiation, described above, documented the coincidence of the onset of scouring with the development of flow instability and standing waves, and wash loads > 35.5 g l^{-1}. In an attempt to provide more precise explanation of the processes involved and particularly to characterize the onset of grain movement, a further set of experiments was carried out (Savat 1979). The conceptual linkage of these experiments is shown in Figure 4.16.

The experiments were carried out in a 4 m long recirculating flume with water supplied from a header tank and in some experiments, from a rainfall simulator. A bed of calcareous loess was used in all experiments, and in some coarse Brusselian sand was superposed. Experiments lasted two hours and with typical discharges used, flows traversed the flume eight times for a maximum 'slope' length of 32 m. Wash loads were found to be about four times lower than predicted from the Yalin bed-load equation. In general total wash load was slightly higher when rills developed. The relationship

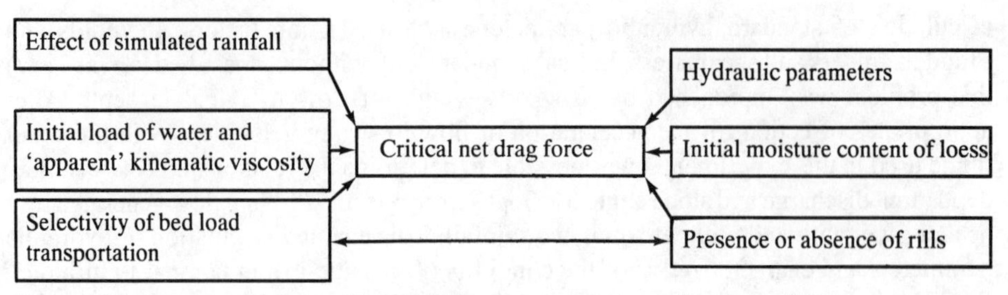

Figure 4.16. Conceptual linkage of Savat's laboratory experiments (after Savat 1979).

between slope angle and rill initiation was found to be erratic and both surface roughness and cohesion appeared to be more significant controls. On smooth surfaces with high cohesion, rills would form only at steep slope angles. Rill initiation was not affected by grooves in the surface or by roughness elements and it was concluded that rill initiation has little to do with vortex erosion. When scouring started, it migrated both up and downslope. Upward migration appeared to be controlled primarily by surface cohesion, and downward migration by the drag coefficient, which increases rapidly with slope steepness and the Froude number. When scouring started on pure loess the average Froude number was 0.81, but locally it reached 1. In experiments with sand over loess, grains started to roll at Fr = 1.2, and complete movement occurred at Fr = 1.6. It appears therefore that rill initiation will always start once the Froude number reaches a critical threshold, which is determined by material characteristics, and close to this threshold vortices play no significant role. As rainfall increases the transport capacity and washload of flow, even without velocity change, rainfall favours rill initiation.

Initially the field applicability of this experiment was restricted because of the size of laboratory installations and the nature of the loessial soils used. It was extended by subsequent field experiments near Leuven and prediction techniques were developed for rill erosion throughout Belgium (Boon and Savat 1980, 1981). It also provided a basis for a significant conceptual study on sheetwash and rill development in which experimental laboratory results were related to the characteristics of rill development reported from many countries (Savat and De Ploey 1982). Rill initiation was found to be related to the very high Froude numbers which occur in sheetwash on steep slopes, which produce supercritical laminar or transitional flow, unlike conditions in deep channels. Critical Froude numbers for rill initiation on Belgian loess, between 2.4-3.0, were shown to occur on slopes of 2-4° steepness. It was demonstrated that Froude number is not dependent on flow discharge, and rill initiation is therefore not constrained by climatic factors, such as the difference between temperate and tropical rainfall, or by slope length. This also explains the occurrence of rill initiation extremely close to drainage divides, particularly on steep slopes. Several other important aspects related to rill initiation were discussed and the need for further study noted. These include the occurrence of flow instability, the development of roll waves and the significance of flow separation in propagation of rills both upslope and downslope through step-scouring. Another aspect requiring study is the relationship between Froude numbers and bedforms beneath the flow. It is pointed out that supercritical flow typically produces undulations in the bed, and that with Froude numbers about 1.2 these will deepen progressively to form rills, even on initially planar surfaces. Finally the importance of rainsplash and the moisture content of the surface soil is discussed. It is shown that in saturated soils of low cohesion rainsplash will cause high sediment transfer from interrill areas and will generate excess pore pressures in rill banks. Together this results in very high sediment loads in rill channels, channel widening, sedimentation, braiding and channel abandonment. Rills developed in such conditions migrate freely across slopes producing more or less uniform denudation and leaving no clearly-defined incised rill system.

4.4.3 *Australian laboratory experiments*

The concepts of surface wash processes and rill initiation evolved by De Ploey, Savat and their associates in Belgium were derived largely from hydraulic laboratory experiments. During approximately the same period Moss and his associates in Australia approached the same point through detailed analysis of sediments collected on hillslopes and in stream channels, together with precise field observations and some laboratory experiments. No transporting sheetflows were observed in the field, all flows being supercritical and turbulent and all flows capable of transportation did so in concentrated channels (Moss and Walker 1978). Clear distinction between suspended load and bed load was found to be possible, even in flows of one millimetre depth. Suspended load was controlled primarily by detachability and virtually no deposition of suspended material was observed on hillslopes. Bedload transport, on the other hand, was governed by flow capacity and was very sensitive to slope gradient. Deposition of a hydraulic mantle of excess bed load occurred on gentler footslopes, a feature regarded as being typical of most natural hillslopes. Within this mantle a clear sequence of bed load deposition could be identified, which was interpreted as being analogous to sequences established for alluvial channel flow, and capable of engendering soil catena development.

Moss and Walker's (1978) study was one of the first to address the linkage between gully and rill development and to show the critical role of bed load capacity. Gullies were defined as narrow, steep-sided channels in areas of net erosion, only partially lined by their own sediments, which flow straight and unite downslope to form larger channels. Rills were defined as occurring in areas of net deposition, with shallow channels cut into, and entirely lined by their own deposits. Downslope they braid and typically split into several smaller channels. While these definitions are by no means universally accepted, by relating 'gully' development on steep hillslopes and 'rill' development on the hydraulic mantle, they do provide a much-needed perspective on the relationship between hillslope channel initiation by surface wash processes and the development of fluvial drainage networks. Gullies developed in heterogeneous regolith typically do not undergo steady, continued incision. Instead a self-limiting process occurs as a coarse channel armour resistant to bed-load transport progressively developed, protecting underlying material vulnerable to suspension transport. If this armour is not breached further gully expansion can take place only by lateral bank erosion, not by incision. The protective armour cannot develop in homogeneous material, and the only limitation to extremely deep incision is then the depth of the regolith itself. This is one factor contributing to the extreme severity of gully erosion in deep loess deposits such as those of Kansu Province in China.

Moss and Walker (1978) also made a valuable contribution through their development of a model for the hydraulic adjustment of mantle slopes. Complete equilibrium adjust-

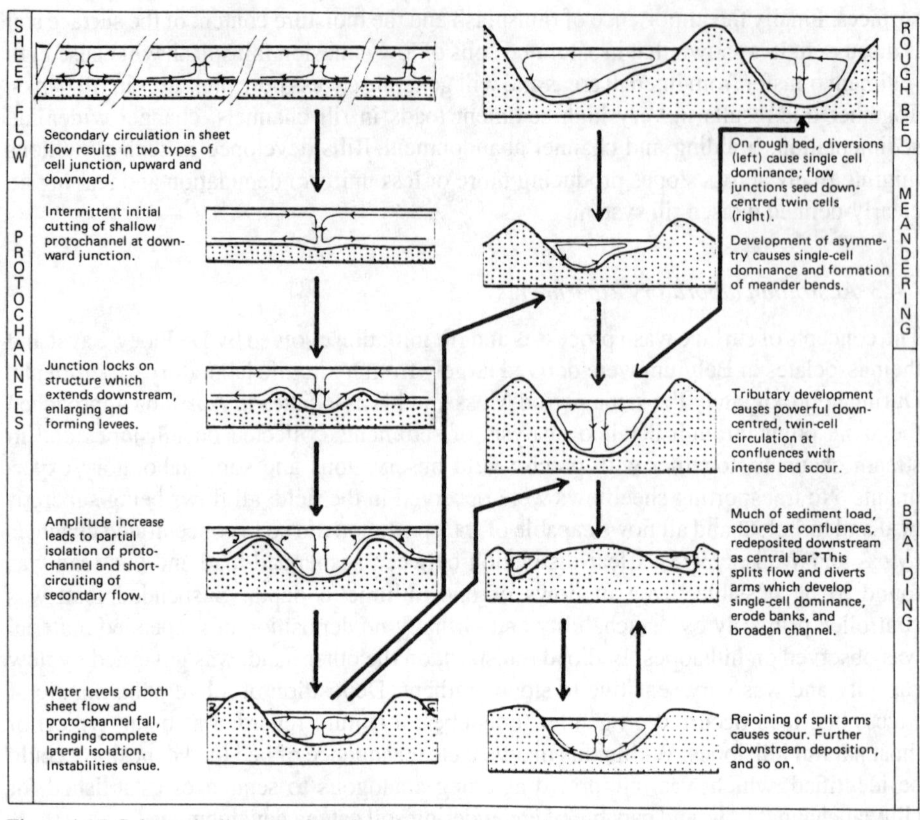

Figure 4.17. Schematic diagram of the dominant secondary circulations involved in the development of proto channels and channels (from Moss et al. 1982).

ment of the slope system can take place only where interaction of water and sediment is not constrained by vegetation, and is therefore most frequently encountered in arid areas. Elsewhere, where vegetation cover is dense, adjustment cannot take place readily and the disequilibrium slope system may be regarded as being in a state of stored hydraulic adjustment potential. Removal of the vegetation cover will release this potential and lead to a period of rapid adjustment until equilibrium is achieved. This model provides a useful framework for relating short-term experimental studies to long-term landscape evolution. If indices of stored hydraulic adjustment potential can be developed it may also ultimately provide a valuable tool for linking short-term field experiments on erosion by surface wash to long-term land-use planning.

This initial partially conceptual study was followed by a series of detailed laboratory experiments with flumes and rainfall simulators designed to clarify the circumstances of rill initiation. All experiments were carried out with artificial flume beds of fine sand. Experiments with flows generated from a header tank and rainfall (Moss et al. 1979) showed that rill channels did not form on low slopes, sheetflow persisting even over minor bed structures. Sediment transport was entirely by rainwash ('rainflow'). On steep slopes rill channels always formed, even if all the water was supplied by rainfall, but on intermediate slopes rill initiation was very sensitive to the precise balance between rainfall and header flow. The suppression of rill channels by rainsplash was attributed to intense small-scale turbulence, flow retardation, and suppression of frictional effects and the development of secondary down-centred flow cells.

In a subsequent study (Moss et al. 1980) found that rill channels will develop spontaneously on an initially plane flume bed. It was strongly suspected that this was linked with down-centred secondary flow cells, but precise observation was prevented by turbidity. This was eliminated in a subsequent study by close control on bed material (Moss et al. 1983). At low slopes and discharges bed armouring frequently occurred and channels did not form, but flow concentration within sheetflow was shown by flotsam lines, each coincident with a down-centred secondary flow. At higher slopes and discharges these flows caused lateral sand grain movement, scouring, and development of 'protochannels', typically about 1 mm deep and 5-10 mm wide. These protochannels develop rather rapidly in a straight line downflow, lateral scouring of sand grains progressively developing small levees on either side of the channel. Eventually these isolate the channel from flow on the surrounding surface, weakening secondary cell circulation and eventually leading to deposition within the protochannel, micro bar formation, meandering and eventually splitting of flow into several secondary cells accompanied by braiding (Fig. 4.17). This stage could also be triggered by tributary inflows to the protochannel.

The protochannel initiation described occurred spontaneously on a plane bed at a critical discharge level. It was also found that protochannels could be 'seeded' beneath this critical discharge by placing small objects on the bed surface. The nature and size of the protochannel formed varied considerably with the size and shape of this object, in some cases two very small protochannels being formed. Analogous channel seeding has also been observed in flume experiments with cohesive beds caused by pebbles and organic fragments spinning and scouring (Bryan, unpubl. data). As detailed hydraulic analysis was not included in Moss et al.'s, (1982) study the results cannot be linked directly to Savat and De Ploey's (1982) experiments in Belgium. However, it was observed that all flows were supercritical and Froude numbers must have been very similar to those observed by Savat and De Ploey to coincide with rill initiation on loess surfaces.

Extension of Moss et al.'s (1982) results to rill initiation on natural hillslopes is still somewhat uncertain. The last set of experiments was carried out without rainfall, which has been shown to suppress secondary flow and therefore any tendency to protochannel formation. The tests also show that protochannels do not form on rough surfaces, though secondary flows do form where flow 'arms' are forced to change direction or where they intersect. Moss et al. (1982) also suggest that in typical heterogeneous regolith coarse material will usually segregate into a stable resistant channel armour which will tend to split the flow into several secondary cells and produce braiding rather than protochannel development.

4.4.4 *Recent developments*

While no linkage has been established between the secondary cell development noted by Walker et al. (1982) and the relationship between rill initiation and flow instability observed by Savat (1976), Karcz and Kersey (1980) have studied the relationship between bedforms and instability in shallow form in a flume experiment at Binghamton. Sediment instability appeared as very small wavelets in the subcritical laminar regime (Re = 55 and Fr = 0.5). Well-developed wave trains appeared at Fr = 1 and persisted throughout the supercritical laminar range yielding to rough flow and disappearing with the transition to turbulent flow at Re around 500. In tests with well-sorted medium sands a marked ridge and trough bed-form appeared at Re above about 55. The precise pattern varied, spacing of ridges sometimes being strikingly regular, and sometimes interspersed by flat spaces. Tests with poorly-sorted fluvio-glacial sands showed marked transverse sorting with coarser particles concentrating progressively under ridges. This would appear to conform to the occurrence of the secondary cells noted by Moss et al. (1982). As flow enters the supercritical laminar regime with Re around 200 the ridge-trough pattern is joined by rhomboidal forms, which become dominant as the supercritical turbulent boundary is reached. While Karcz and Kersey (1980) identified numerous aspects of instability forms in shallow flow which require further study, they noted that the experimental data support Horton's (1938) reasoning on the potential erosional importance of 'rain wave trains' albeit starting well within the laminar sub-critical regime. They also noted the potential applicability to development of natural rills.

Considerable further information on the effect of sediment heterogeneity and armouring noted by Moss et al. (1982) has been contributed by Rowntree's (1982) experiments in Scotland, carried out with a 6 × 3 m laboratory flume. In these flow was generated entirely by simulated rainfall but kinetic energy was so low that rainsplash influence was minimal. Three experiments were carried out, each consisting of a sequence of six storms of varied duration. In the first homogeneous fine sands very similar to those used by Moss et al. (1982) were used, while in the other coarser and more heterogeneous material was used. Detailed observations were made of changes in sediment yield and progressive development of rill networks during storm sequences. By far the highest sediment yields were observed in the experiment with homogeneous sand. Although yields varied over time no pronounced trend could be discerned and no armouring was observed. A well-integrated rill network progressively developed which eventually covered the complete slope. In experiments with other materials, by contrast, clear evidence of armouring was observed, particularly on the most heterogeneous material (D_{50} = 0.3 mm). On this material a stable rill network formed rapidly which did not expand significantly after the second storm. This allowed armouring on the surface and channel to develop its maximum effect, reflected in very low and progressively declining sediment yield. Rill network development on the coarsest material (D_{50} = 1.0 mm) was less rapid but expansion continued throughout the storm sequence. The effects of armouring on sediment yield only became apparent close to the end of the experiment, the earlier effect of armour development on channel beds being compensated for by progressive exposure of erodible channel wall material in the expanding rill network. A particularly interesting component of this study was the use of erodents to identify the source areas of sediment. In all three experiments a progressive shift from rill to interrill source areas was noted during the storm sequence.

4.4.5 *Summary remarks*

The laboratory studies in Belgium, Australia and Scotland, described above, have approached the problem and significance of the relationship between surface wash and rill initiation from quite different directions. Collectively they have provided much information about the circumstances which can lead to rill initiation, but there is still a critical need for further study of numerous aspects of the processes involved. Some of the more significant are listed below:

a) All the studies described have used cohesionless or low-cohesion materials. This was essential to simplify experiments in a first approximation, but it is clear that general application of the results will require experiments on cohesive materials, and particularly on naturally-aggregated soils, where subtle variations in antecedent moisture conditions may dramatically influence material behaviour. This has been observed in laboratory flume experiments with natural cohesive soils (Bryan, unpubl. data) and in field experiments on swelling, montmorillonite-rich shales (Bowyer-Bower and Bryan 1984).

b) Further experiments are necessary to determine if the effects observed by Moss et al. (1982) and Rowntree (1982) occur under full rainfall energy conditions. Several of the studies have indicated the suppression of channel-forming tendencies by rainsplash impact. Dunne (1980) has also observed that transfer of sediment by rainsplash from exposed interrill areas to adjacent channels may prevent rill expression or preservation. This effect can occur, but its significance must vary with the recessional characteristics of overland flow after rainstorm. Hodges (1982) has also shown the obliteration of scoured channels by recessional deposition unrelated to transfer from interrills. Dunne (1980) has also pointed out that rainsplash will also tend to counteract rill incision because it generally occurs much more frequently, in storms which do not generate overland flow.

c) Most of the studies described deal only with surface processes. On many hillslopes, particularly on clay-rich materials, rill initiation is influenced by subsurface water flow and micropiping. This has been shown by field studies in badland areas (Bryan et al. 1978, Yair et al. 1980, Bowyer-Bower and Bryan 1984), where physical erosion processes are well-developed, but should also be examined on true soils where micro-faunal activity and biochemical processes are also significant. Virtually no attention has been paid to the importance of chemical and solution processes in rill initiation apart from the experimental studies of rillenkarren formation on limestone blocks and on limestone in the field and on plaster of paris blocks in the laboratory by Glew and Ford (1980).

d) Further experiments should be carried out on the range and potential significance of instability features in shallow flow in the field and laboratory, following the observations of Savat (1976) and Karcz and Kersey (1980). It is particularly important that the existence of these features under rainsplash impact be examined and their relationships with such variables as slope configuration be established. Experiments on these aspects are currently in progress with a new 25 m flume at Toronto in which numerous slope configurations can be tested. Related experiments on the hydraulics of sediment transported by overland flow in relation to changes in Fr and Re and rill initiation are also in progress at Silsoe, Bedfordshire by Morgan and his associates.

4.5 RAINSPLASH EXPERIMENTS

Several of the comments above emphasize the need for further experimental research on the interaction between rainsplash, surface wash and rill initiation. Space does not permit detailed comments on all recent research or experiments currently in progress on rainsplash, but brief reference will be made to some more salient work. Rainsplash has received particular attention recently from geomorphologists in Belgium. Bollinne (1978, 1980, 1982) has emphasized field measurement of splash rates over a prolonged period on the loessial soils of Hesbaye in central Belgium. The results show a complex interrelationship between splash and wash loss and soil properties, splash loss being positively correlated with soil structural indices, while wash loss showed negative correlation.

Most of the experiments on rainsplash have involved vertical rain without wind disturbance. Moeyersons (1983) has reported the results of experiments carried out in Rwanda which assessed the effect of oblique wind-driven rain. It was found that, as in Lyles' (1976) earlier study, wind direction and strength strongly influence the amount and direction of splash transport.

Poesen and Savat also carried out field splash studies near Leuven, but the main focus of their work has been laboratory analysis of splash processes. This has been particularly notable for accurate assessment of splash transport using a new weighing technique (Savat 1981), whereas most rainsplash studies have assessed only splash detachment. Precise functional relationships were established between splash detachability, splash transportability and sediment size, expressed as the threshold kinetic energy necessary to move particles of given size (Savat and Poesen 1981, Poesen and Savat 1981). Detachability was positively correlated with soil water content, but also showed time-dependent variations related to particle size and the evolution of a resistant 'pavement'. Contrary to the results of some earlier studies, no protective effect was found to result from the development of a surface water layer (Poesen 1981). Transportability was not affected by soil water content but did show a time-dependent decrease due to influence of a developing surface water layer on splash ejection angles. Poesen subsequently expanded the scope of this research with more extensive laboratory experiments and field observations at the Huldenberg research site near Leuven (Poesen 1983). The results have been incorporated with many data from other parts of the world in an improved splash transport model (Poesen 1985). Apart from providing extensive new data on rainsplash processes this work also provides an excellent review of recent experimental research on rainsplash, including that of Morgan (1982) and Quansah (1981) in England, McCarthy (1980) in the United States, and Reeve (1982) in Australia.

4.6 CONCLUSIONS

Despite recent advances in a number of areas it is impossible to avoid the conclusion that our current understanding of the nature and significance of surface wash processes is in no way commensurate with the vast amount of effort which has been devoted to experiments and monitoring studies. The reasons are manifold, but it seems clear that the opportunity to acquire an excellent data base for theoretical analysis has been largely missed through failure to standardize methods or even to develop a common understanding of the sort of information which should be collected.

The problem is most clearly apparent when runoff plot studies are considered which have been, and are, the most common types of surface wash experiments. It is well-known that hydrological response and sediment yield are markedly affected by the size, shape and steepness of the drainage area. Despite this common knowledge, almost every conceivable size and shape of runoff plot has been employed in erosion studies. As a result the wealth of data produced can be applied only to the site of origin and cannot be usefully incorporated in an attempt to move from empirically to physical-based models of soil erosion and hillslope development.

The lack of comparability resulting from failure to standardize runoff plot characteristics is exacerbated by the frequent inadequacy of information by which the results reported can be interpreted. This relates to many aspects of experimental design but only the most important will be discussed. Antecedent soil moisture conditions have been shown in many studies to be a critical control on the timing and magnitude of runoff generation and sediment transport during storm events, yet there are very few runoff plot experiments in which any systematic attempt has been made to monitor this regularly or to correct results. Another problem is the identification of the area of the plot contributing runoff and sediment, despite the clear recognition that partial area and variable source area hydrological concepts frequently apply even at very small scales. While it is sometimes difficult to identify source areas clearly during storm events, it is at least highly desirable that sediment transport data reported on a weight-per-unit area basis be correlated with some information on the distribution of denudation areas over the plot area.

Perhaps the single most critical gap is the inadequacy of information about rainfall. Many experiments are reported without information on rainfall intensity, and sometimes with rainfall amounts reported only on a monthly or annual basis. By itself this is totally inadequate for detailed interpretation of surface wash processes or their relationship to controlling variables.

One point that emerges clearly from the preceding discussion is that detailed comprehension of surface wash processes cannot be derived from aggregated data. It is essential that observations be carried out and reported on an individual storm basis. This does not mean that long-term observations which provide some perspective on seasonal or longer-term variations are not also necessary, but without data from individual storms precision of data analysis must be most inadequate. One useful technique for provision of detailed unit storm information without high expenditures or monitoring equipment or personnel is rainfall simulation. Although this has been widely used and has produced valuable results, again the full possibilities for data comparison and integration have been missed through lack of standardization. Without espousing any particular simulator it is necessary to emphasize that unless reproduction of natural rainstorm conditions is exact, only results produced with identical simulators will be comparable. In view of the clear imperfections of most simulators in common use it is reasonable to suggest that more advantages can be achieved by standardized use of cheap, portable and readily available simulators, even if they are imperfect, rather than an elusive quest for perfect rainfall reproduction.

The lack of experimental precision apparent in many runoff plot studies has generally been avoided in experimental microplot and laboratory studies and it is from these that most of the detailed knowledge of surface wash processes has emerged. Considerable progress has been achieved in the past fifteen years but major gaps still exist which require careful experimentation. Despite a number of excellent field studies too many of the available data are derived from laboratory studies, and field applicability is not always clear. Of necessity most of the laboratory studies have used artificial cohesionless sediment mixtures, or natural soils in which field structure has been largely or completely obliterated. While there has been little comparative laboratory and field testing using standard methodology, it does appear that relationships based on laboratory data can be applied with some confidence to natural soils of low cohesion and to poorly structured soils which have been disturbed repeatedly by agricultural activity. It is much less certain that they can be applied with equal success to highly developed, cohesive, vegetated and often stoney soils usually encountered on hillslopes where surface wash processes most often dominate in nature. If a sound understanding of the geomorphic and hydrologic significance of surface wash is to be developed extensive experimentation on undisturbed natural soil profiles in the field would appear to be essential. This should be accompanied by considerably more attention in laboratory experiments to the particular problems and behaviour of cohesive soils.

In carrying out extended field and laboratory experiments on natural soils it is desirable that the research perspective be broader than in most previous studies. Experimental design should encompass detailed hydraulic observations, assessment of temporal and spatial variations in material sedimentology and time-dependent changes in process distribution. In particular future experiments should be designed to incorporate recent information on rainsplash processes with the full range of surface wash phenomena. In connection with process distribution one major weakness in available data at present is inadequate information on the recurrence interval and magnitude of surface wash on many hillslopes. At present only the studies of Pearce (1976) and Dunne and Dietrich (1980a) have addressed this question explicitly, and a much more extensive data base is desirable.

Experimental design for both laboratory and field experiments should recognize more explicitly the interrelationship of surface wash and pedogenic processes. This means that experiments should include not only physical processes of runoff generation, entrainment and transport, but should also examine carefully the role of biological activity on water movement and material resistance. This is particularly important where time-dependent crusting phenomena limit sediment availability. More attention should also be paid to the chemical characteristics of soils, as they affect physical dispersion and erodibility and also the removal of material in solution. Studies of the interaction of pedogenic and surface wash processes should also recognize the role of the latter in generating diverse soil types, and in particular the development of soil catenas. At present this has been formally studied only by the experiments of Yair and his associates in Israel and Moss and his associates in Australia. In designing new experiments it is important to recognize the extreme variability of natural soil systems and field study only provides a framework for interpreting results where replication has not taken place. It is necessary to go further and to establish experimental replication as a standard component of surface wash research, as it is in other experimental sciences, despite the difficulty, cost and inconvenience involved. It is possible that some of the interactions observed experimentally are not reproducible. For example, this possibility has been discussed in connection with the behaviour of montmorillonite-rich shales in the Alberta badlands of Canada in response to wetting. It would be useful to examine this question more formally in future surface wash experiments.

Beyond the desirability of replication and standardization of methodology and instrumentation it is difficult to make general recommendations for experimental design. It is clear that experiments should be carried out in the widest possible range of environmental conditions. It is important that detailed experimental research be linked to an integrated appraisal of landscape evolution on a regional scale. In the case of surface wash this will require close attention to the linkages between hillslopes and drainage system integration, and the interaction with other geomorphic processes. This may be achieved both by careful field research and by further development of process-based mathematical models such as those introduced by Band (1983). Finally it is necessary to develop a clear perspective on the temporal significance of the processes being studied. One possible framework for this is the concept of 'stored hydraulic adjustment potential' introduced by Moss and Walker (1978). Such a framework will provide a channel for the integration of experimental field and laboratory research and regional geomorphic research into the preparation of land-use planning recommendations. This will certainly improve the quality of land-use planning, but it seems also the most likely source of financial and institutional support for the extended programmes of experimental research essential for progress in theoretical geomorphology.

REFERENCES

Aina, P.O., Lal, R. and Taylor, G.S., 1977. Soil and crop management in relation to soil erosion in the rainforest of western Nigeria. *Soil erosion: prediction and control*, Soil Conservation Soc. Amer., Special Publication 21, 75-82.

Asseline, J., and Valentin, C., 1978. Construction et mise au point d'un infiltrometre à aspersion. *Cahiers ORSTOM*, séries Hydrology, 15, 321-349.

Aubry, B. and Wahome, E.K., 1983. Field experiments on soil erosion in Amboseli, Kajiado District. In D.B. Thomas and W.M. Senega (eds.), *Soil and water conservation in Kenya*, Institute for Development Studies and Faculty of Agriculture, University of Nairobi, 65-78.

Band, L.E., 1985. Simulation of slope development and the magnitude and frequency of overland flow erosion in an abandoned hydraulic mine pit. In M.J. Woldenberg (ed.), *Models in geomorphology*, George Allen and Unwin, 191-211.

Band, L.E., 1983. Measurement and simulation of hillslope development,Unpublished Ph.D. thesis, University of California, Los Angeles.

Barber, R.G., 1983. The magnitude and sources of soil erosion in some humid and semi-arid parts of Kenya and the significance of soil-loss tolerance values in soil conservation in Kenya. In D.B. Thomas and W.M. Senega (eds.), *Soil and water conservation in Kenya*, Institute for Development Studies and Faculty of Agriculture, University of Nairobi, 20-47.

Barber, R.G., Moore, T.B., and Thomas, D.B., 1979. The erodibility of two soils from Kenya. *J. Soil Science*, 30, 579-591.

Barber, R.G., Thomas, D.B., and Moore, T.R., 1981. Studies on soil erosion and runoff and proposed design procedures for terraces in the cultivated, semi-arid areas of Machakos District, Kenya. In R.P.C. Morgan (ed.), *Soil conservation: problems and prospects,* Wiley-Interscience, Chichester, 219-237.

Bazzoffi, P., Torri, D. and Zanchi, C., 1980. Estimate of soil erodibility by means of simulated rain in laboratory, I. Rainfall Simulator, Instituto Sperimentale per lo studio e la difesa del Suolo – Firenze, Report.

Becher, H.H., 1981. Auswirkungen der profiver kurzungen durch wassererosion auf den ertrag, *H:H. DBG*, 30, 341-342.

Binard, M. and Bollinne, A., 1980. Contribution à l'étude quantitative des modifications des risques d'érosion résultant des rémembrements, *Pédologie*, 30, 323-333.

Blackie, J.R., Edwards, K.A., Clarke, R.T., 1979. Hydrological research in East Africa, *East African Agricultural and Forestry Journal*, 43. (Special Edition).

Bollinne, A., 1977. La vitesse de l'érosion sans culture en region limoneuse, *Pédologie*, 27, 191-206.

Bollinne, A., 1978. Study of the importance of splash and wash on cultivated loamy soils of Hesbaye (Belgium). *Earth Surface Processes*, 3, 71-84.

Bollinne, A., 1980. Splash measurements in the field. In D. Gabriels and M. De Boodt (eds.), *Assessment of soil erosion*, Wiley-Interscience, Chichester, 441-453.

Bollinne, A., 1982. Etude et prévision de l'érosion des sols limoneux cultivés en Moyenne Belgique, Unpublished DSc géog, thesis, Université de Liége.

Bollinne, A., and Rosseau, P., 1978. L'érodibilité des sols de Moyenne et Haute Belgique, *Bull. Soc. Géog. Liége*, 14, 127-140.

Boon, W. and Savat, J., 1980. Computer mapping of rill erosion in the Belgian Scheldt basin, *Pédologie*, 30.

Boon, W. and Savat, J., 1981. A nomogram for the prediction of rill erosion. In R.P.C. Morgan (ed.), *Soil conservation: problems and prospects.* Wiley-Interscience, Chichester, 303-319.

Bork, H.R., and Rohdenburg, H., 1979. The behaviour of overland flow and infiltration under simulated rainfall, in *Seminar on agricultural soil erosion in temperate non-mediterranean climate*, Universite Louis-Pasteur, Strasbourg, 225-237.

Bork, H.R. and Rohdenburg, H., 1980. Rainfall simulation in south-east Spain: analysis of overland flow and infiltration. In M. De Boodt and D. Gabriels (eds.), *Assessment of soil erosion*, Wiley-Interscience, Chichester, 293-302.

Bowyer-Bower, T.A.S., 1984. A study of rills in a natural landscape, Unpublished M.Sc. thesis, University of Toronto.

Bowyer-Bower, T.A.S. and Bryan, R.B., 1984. Rill initiation: concepts and experimental evaluation on badland slopes. Paper presented to the *Symposium of the I.G.U. commission on field experiments in geomorphology*, Strasbourg.

Bryan, R.B., 1973. Surface crusts formed under simulated rainfall on Canadian soils, *C.N.R. laboratorio per la chimica del terreno, Pisa*, Conferenze 2, 30 pp.

Bryan, R.B., 1981. Soil erosion under simulated rainfall in the field and the laboratory: variability of erosion under controlled conditions. *Erosion and sediment transport measurement*, IAHS Public., 133, Florence, 391-403.

Bryan, R.B. and Campbell, I.A., 1986. Runoff and sediment discharge in a semi-arid ephemeral drainage basin. *Z. Geomorph. Supp. Bd.* 58, 121-143.

Bryan, R.B. and De Ploey, J., 1983. Comparability of soil erosion measurements with different rainfall simulators. In J. De Ploey (ed.), *Rainfall simulation, runoff and soil erosion*, Catena Supp. Bd., 4, 33-56.

Bryan, R.B. and Harvey, L.E., 1985. Observations on the geomorphic significance of tunnel erosion in a semi-arid ephemeral drainage system. *Geografiska Annaler* 67A, 257-272.

Bryan, R.B., Imeson, A.C., and Campbell, I.A., 1984. Solute release and sediment entrainment on microcatchments in the Dinosaur Park Badlands, Alberta, Canada, *J. Hydrology*, 71, 79-106.

Bryan, R.B. and Luk, S.-H., 1981. Laboratory experiments on the variation of soil erosion under simulated rainfall, *Geoderma*, 26, 245-265.

Bryan, R.B., Yair, A. and Hodges, W.K., 1978. Factors controlling the initiation of runoff and piping in Dinosaur Provincial Park badlands, Alberta, Canada. *Z. Geomorph. Supp. Bd.*, 29, 151-168.

Campbell, I.A., 1982. Surface morphology and rates of change during a ten-year period in the Alberta badlands. In R.B. Bryan and A. Yair (eds.), *Badland geomorphology and piping*, Geo Books, Norwich, 221-237.

Carson, M.A. and Kirkby, M.J., 1972. *Hillslope form and process*. Cambridge University Press.

Chisci, G. and D'Egidio, G., 1981. Erosion investigation-watersheds, hydrology and sediment yield modelling on small watersheds. *Erosion and sediment transport measurement symposium*, Field Excursion Guide, IASH, Florence, 38-40.

Chisci, G. and Zanchi, C., 1981. The influence of different tillage systems and different crops on soil losses on hilly silty-clayey soils. In R.P.C. Morgan (ed.), *Soil conservation problems and prospects*, Wiley-Interscience, Chichester, 210-217.

Chisci, G., Zanchi, C. and D'Egidio, G., 1981. Erosion investigation – plots. *Erosion and sediment transport measurement symposium*, Field Excursion Guide, IASH, Florence, 40-46.

Collinet, J. and Valentin, C., 1982. Effects of rainfall intensity and soil surface heterogeneity on steady infiltration rate, *12th international congress soil science, New Delhi*.

Dediu, M. and Cazangiu, F., 1983. Utilisation des simulateurs de la pluie dans les expérimentations géomorphologiques pour l'étude de l'aménagement des terrains agricoles, Paper to Symposium on *The role of geomorphological field experiments in land and water management*, Bucharest.

De Ploey, J., 1969. L'érosion pluviale: expériences à l'aide de sables traceurs et bilans morphogéniques, *Acta Geographica Lovaniensia*, 7, 1-28.

De Ploey, J., 1977. Some experimental data on slope wash and wind action with reference to Quaternary morphogenesis in Belgium, *Earth Surface Processes*, 2-3, 101-106.

De Ploey, J., 1979. A consistency index and the prediction of surface crusting on Belgian loamy soils. *Seminar on agricultural soil erosion in temperate non-mediterranean climates*, University of Strasbourg, 133-137.

De Ploey, J., 1981. Crusting and time-dependent rainwash mechanisms on loamy soil. In R.P.C. Morgan (ed.), *Soil conservation: problems and prospects.* Wiley-Interscience, Chichester, 139-154.

De Ploey, J. and Moeyersons, J., 1975. Runoff creep of coarse debris: experimental data and some field observations. *Catena*, 2, 275-288

De Ploey, J. and Mucher, H.J., 1981. A consistency index and rainwash mechanisms on Belgian loamy soils. *Earth Surface Processes and Landforms*, 6, 319-330.

De Ploey, J. and Savat, J., 1968. Contributions à l'étude de l'érosion par le splash, *Z. Geomorph.*, 2, 174-193.

De Ploey, J., Savat, J. and Moeyersons, J., 1976. The differential impact of some soil factors on flow, runoff creep and rainwash, *Earth Surface Processes*, 1, 151-161.

DeVleeschauwer, D., Lal, R. and De Boodt, M., 1978. Comparison of detachability indices in relation to soil erodibility for some important Nigerian soils, *Pédologie*, 28, 5-20.

Djorovic, M., 1980. Slope effect on runoff and erosion. In M. De Boodt and D. Gabriels (eds.), *Assessment of erosion*, Wiley-Interscience, Chichester, 215-225.

Dunne, T., 1980. Formation and controls of channel networks. *Progress in Physical Geog.*, 4, 211-239.

Dunne, T. and Dietrich, W.E., 1980a. Experimental study of Horton overland flow on tropical hillslopes, I. Soil conditions, infiltration and frequency of runoff, *Z. Geomorph., Supp. Bd.*, 35, 40-59.

Dunne, T. and Dietrich, W.E., 1980b. Experimental investigation of Horton overland flow on tropical hillslopes, 2. Hydraulic characteristics and hillslope hydrographs. *Z. Geomorph., Supp. Bd.*, 35,60-80.

Dunne, T., Dietrich, W.E. and Brunengo, M.J., 1980. Simple, portable equipment for erosion experiments under artificial rainfall, *J. Agric. Engineering Research*, 25, 161-168.

El-Swaify, S.A., 1977. Susceptibilities of certain tropical soils to erosion by water. In D.J. Greenland and R. Lal (eds.), *Soil conservation and management in the humid tropics*, Wiley-Interscience, Chichester, 70-77.

El-Swaify, S.A., and Dangler, E.W., 1976. Erodibilities of selected tropical soils in relation to structural and hydrologic parameters. *Soil erosion prediction and control*, Soil Conservation Society of America, Special Publication 21, 105-114.

Elwell, H.A., 1977. A soil loss estimation system for southern Africa, *Rhodesia Department Agric. Research Bull.* 22.

Emmett, W.W., 1970. The hydraulics of overland flow on hillslopes, *US Geol. Surv. Prof. Paper*, 662-A.

Engelund, F., 1967. Hydraulic resistance of alluvial channels: closure of discussion. *J. Hydraulics Div.*, ASCE, 93 (HY 4), 287-296.

Epema, G.F. and Riezebos, H.T., 1983. Fall velocity of waterdrops at different heights as a factor influencing erosivity of simulated rain, *Catena Supp.*, 4, 1-17.

Evenari, M., Shanan, L. and Tadmor, H., 1968. Runoff farming in the desert. I. Experimental layout. *Agronomy J.*, 60, 29-32.

Farres, P., 1978. The role of time and aggregate size in the crusting process, *Earth Surface Processes*, 3, 243-254.

Froehlich, W., 1982. The mechanism of fluvial transport and waste supply into the stream channel in a mountainous Flysch Catchment, *Polish Academy of Sciences, Geographical Studies*, 143.

Gabriels, D., Pauwels, J.M. and De Boodt, M., 1977. A quantitative rill erosion study on a loamy sand in the hilly region of Flanders, *Earth Surface Processes*, 2-3, 257-260.

Gerlach, T., 1979. Action des gouttes de pluies, *Sem. agric. soil erosion in temperate non-mediterranean climate*, Strasbourg-Colmar, 61.

Giordano, A. and Mattioli, L., 1981. Rainfall simulation for the quantitative evaluation of erosion effect on different soils. *Erosion and sediment transport measurement symposium*, Field Excursion Guide, IASH, Florence, 19-27.

Glew, J.R. and Ford, D.C., 1980. A simulation study of the development of rillenkarren, *Earth Surface Processes*, 5, 25-36.

Hamilton, H., 1984. An experimental investigation into the influence of antecedent soil moisture content on soil erodibility, Unpublished M.Sc. thesis, University of Toronto.

Hatch, T., 1981. Preliminary results of soil erosion and conservation trials under pepper (Piper nigrum) in Sarawak, Malaysia. In R.P.C. Morgan (ed.), *Soil conservation: problems and prospects.* Wiley-Interscience, Chichester, 255-262.

Hillel, D., 1960. Crust formation in loessial soils. *Proc. 7th international congress of soil science*, 330-339.

Hodges, W.K., 1982. Hydraulic characteristics of a badland pseudo-pediment slope system during simulated rainstorm experiments. In R.P. Bryan and A. Yair (eds.), *Badland geomorphology and piping*, Geo Books, Norwich, 127-151.

Hodges, W.K., 1984. Experimental study of hydrogeomorphological processes in Dinosaur Badlands, Alberta, Canada. Unpublished Ph.D. thesis, University of Toronto.

Hodges, W.K. and Bryan, R.B., 1982. The influence of material behaviour on runoff initiation in the Dinosaur Badlands, Canada. In R.B. Bryan and A. Yair (eds.), *Badland geomorphology and piping*, Geo Books, Norwich, 13-46.

Horton, R.E., 1938. Rainwave trains. *Trans. Amer. Geophysical Union*, 19, 367-374.

Hudson, N.W., 1957. The design of field experiments on soil erosion. *J. Agric. Engineering Research*, 2, 56-65.

Hudson, N.W., 1981. Instrumentation for studies of the erosive power of rainfall. *Erosion and sediment transport measurement*, IAHS Publication, 133, Florence.

Imeson, A.C., 1976. Some effects of burrowing animals on slope processes in the Luxembourg Ardennes, *Geografiska Annaler*, 58, Part I, 115-125, Part II, 317-328.

Imeson, A.C., 1983. Studies of erosion thresholds in semi-arid areas: field measurements of soil loss and infiltration in northern Morocco. *Catena Supp.*, 4, 79-89.

Imeson, A.C, and Jungerius, P.D., 1976. Aggregate stability and colluviation in the Luxembourg Ardennes; and experimental and micromorphological study, *Earth Surface Processes*, 1, 259-271.

Imeson, A.C., Kwaad, F.J.P.M. and Verstraten, J.M., 1982. The relationship of soil physical and chemical properties to the development of badlands in Morocco. In R.B. Bryan and A. Yair (eds.), *Badland geomorphology and piping*, Geo Books, Norwich, 47-70.

Imeson, A.C., Vis, R. and De Water, E., 1981. The measurement of water-drop impact forces with a piezo-electric transducer. *Catena*, 8, 83-96.

Ionita, I., 1983. The Perieni-Birlad Central Research Station for soil erosion control, *Field excursion guide-book for symposium on 'The role of geomorphological field experiments in land and water management'*, Bucharest.

Iwagaki, Y., 1955. Fundamental study on the mechanisms of soil erosion by overland flow. Unpublished Ph.D. thesis, Kyoto University (in Japanese).

Izzard, R.F., 1944. The surface profile of overland flow. *Trans. Amer. Geophysical Union*, 25, 959-968.

Izzard, R.F., 1946. Hydraulics of runoff from developed surfaces, *Proc. 26th annual meeting*, Highway Research Board, National Research Council, National Academy of Sciences, Washington, D.C.

Johnston, H.T., Elsawy, E.M. and Cochrane, S.R., 1980. A study of the infiltration characteristics of undisturbed soil under simulated rainfall, *Earth Surface Processes*, 5, 159-174.

Kamphorst, A. and Bolt, G.H., 1976. Saline and sodic soils. In G.H. Bolt and M.G.M. Bruggenwert (eds.), *Soil chemistry: basic elements*, Elsevier, Amsterdam, 171-191.

Karcz, I. and Kersey, D., 1980. Experimental study of free-surface flow instability and bedforms in shallow flows, *Sedimentary Geol.*, 27, 263-300.

Kerenyi, A. and Pinczes, Z., 1979. Methods for and results on the examination of the effect of baserock and slope angle on erosion. *Seminar on agricultural soil erosion in temperate non-mediterranean climate*, Université Louis Pasteur, Strasbourg, Institut National de la Recherche Agronomique, Colmar, 81-87.

Kinnell, P.I.A., 1976. Some observations of the Joss-Waldvogel rainfall distrometer, *J. Applied Meteorol.*, 15, 499-502.

Kowal, J.M. and Kassam, A.H., 1977. Energy load and instantaneous intensity of rainstorms at Samaru, Northern Nigeria. In D.J. Greenland and R. Lal (eds.), *Soil conservation and management in the humid tropics*, Wiley-Interscience, Chichester, 57-70.

Lal, R., 1976a. Soil erosion problems on an Alfisol in western Nigeria and their control, *IITA Monograph I.*

Lal, R., 1976b. Soil erosion on Alfisols in Western Nigeria I. Effects of slope crop rotation and residue management, *Geoderma*, 16, 363-375.

Lal, R., 1976c. Soil erosion on Alfisols in Western Nigeria II. Effect of mulch rate, *Geoderma*, 16, 377-387.

Lal, R., 1976d. Soil erosion on Alfisols in Western Nigeria III. Effect of rainfall characteristics, *Geoderma*, 16, 389-401.

Lal, R., 1976e. Soil eroson on Alfisols in Western Nigeria IV. Nutrient losses in eroded sediments, *Geoderma*, 16, 403-417.

Lal, R., 1976f. Soil erosion on Alfisols in Western Nigeria V. Changes in soil physical characteristics and crop response, *Geoderma*, 16, 419-431.

Lal, R, 1977. A brief review of erosion research in the tropics of south-east Asia. In D.J. Greenland and R. Lal (eds.), *Soil conservation and management in the humid tropics*, Wiley-Interscience, Chichester, 203-212.

Lal, R., 1980. Soil erosion problems on Alfisols in western Nigeria VI. Effects of erosion on experimental plots, *Geoderma*, 25, 215-230.

Lal, R., 1982. Effects of slope length and terracing on runoff and erosion on a tropical soil, pp. 23-31. *Recent developments in the explanation and prediction of erosion and sediment yield*, IAHS Publication, 137, 23-31.

Lal, R., 1983. Effects of slope length on runoff from Alfisols in Western Nigeria, *Geoderma*, 28.

Laronne, J.B., 1981. Dissolution kinetics of surficial Mancos Shale-associated alluvium, *Earth Surface Processes and Landforms*, 6, 541-552.

Laronne, J.B., 1982. Sediment and solute yield from Mancos Shale hillslopes, Colorado and Utah. In R.B. Bryan and A. Yair (eds.), *Badland geomorphology and piping*, Geo Books, Norwich, 181-193.

Laronne, J.B. and Shen, H.W., 1982. The effect of erosion on solute pickup from Mancos Shale hillslopes, Colorado, USA, *J. Hydrology*, 59, 189-207.

Leser, H., 1980. Soil erosion measurements on arable land in north-west Switzerland. *Geography in Switzerland*, Berne, I.G.U. Tokyo, 9-14.

Leser, H., Schmidt, R-G. and Seiler, W., 1981. Bodenerosion in Hochrheintal und Jura (Schweiz), *Petermanns Geographische Mitteilungen*, 2/81, 83-91.

Lowdermilk, W.C. and Sundling, H.L., 1950. Erosion pavement formation and significance, *Trans. Amer. Geophysical Union*, 31, 96-106.

Luk, S-H., 1982. Variability of rainwash erosion within small sample areas. In C. Thorn (ed.), *Space and time in geomorphology*, George Allen and Unwin, London, 243-268.

Luk, S-H., 1983. Effect of aggregate size and microtopography on rainwash and rainsplash erosion, *Z. Geomorph.*, 27, 283-295.

Luk, S-H. and Morgan, C., 1981. Spatial variations of rainwash and runoff within apparently homogeneous areas, *Catena*, 8, 383-402.

Lyles, L., 1977. Soil detachment and aggregate disintegration by wind-driven rain. *Soil erosion prediction and control*, Soil Conservation Society of America, Special Publication, 21, 152-159.

McCarthy, C., 1980. Sediment transport by rainsplash. Unpublished Ph.D. thesis, University of Washington.
Messer, T., 1980. Soil erosion measurements on experimental plots in Alsace vineyards (France). In M. De Boodt and D. Gabriels (eds.), *Assessment of erosion*, Wiley-Interscience, Chichester, 455-462.
Moeyersons, J., 1983. Measurements of splash-saltation fluxes under oblique rain, *Catena*, Supp. Bd. 4, 19-31.
Moldenhauer, W.C. and Foster, G.R., 1981. Empirical studies of soil conservation techniques and design procedures. In R.P.C. Morgan (ed.), *Soil conservation: problems and prospects*, Wiley-Interscience, Chichester, 13-29.
Moore, T.R., Thomes, D.B. and Barber, R.G., 1979. The influence of grass cover on runoff and soil erosion from soils in the Machakos area, Kenya, *Tropical Agric.* (Trinidad), 56, 339-344.
Morin, J., Goldberg, D. and Seginer, I., 1967. A rainfall simulator with a rotating disc, *Trans. Amer. Soc. Agric. Engineers*, 10, 74-77.
Morand, F., 1979. Les parcelles de mesure des processus d'érosion actuels du Mont des Vaux (Cessières-Aisne, France), *Agric. soil erosion in a temperate non-mediterranean climate*, Strasbourg-Colmar, 73-87.
Morgan, C., 1979. Field and laboratory examination of soil erosion as a function of erosivity and erodibility for selected hillslope soils from Southern Ontario, Unpublished Ph.D. Thesis, University of Toronto.
Morgan, R.P.C., 1978. Recherches sur l'erosion des sols sableux en Bedfordshire, Angleterre, *Agricultural soil erosion in temperate non-mediterranean climates*, Universite Louis Pasteur, Strasbourg, 35-38.
Morgan, R.P.C., 1978. Field studies of rainsplash erosion. *Earth Surface Processes*, 3, 295-299.
Morgan, R.P.C., 1980. Field studies of sediment transport by overland flow, *Earth Surface Processes*, 5, 307-316.
Morgan, R.P.C., 1982. Splash detachment under plant covers: results and implications of a field study, *Trans. Amer. Soc. Agric. Engineers*, 25,987-991.
Moss, A.J., Green, P. and Hutka, J., 1982. Small channels: their experimental formation, nature and significance. *Earth Surface Processes and Landforms*, 7, 401-416.
Moss, A.J. and Walker, P.H., 1978. Particle transport by water flows in relation to erosion, deposition, soils and human activities, *Sedimentary Geol.*, 20, 81-139.
Moss, A.J., Walker, P.H. and Hutka, J., 1980a. Movement of loose, sandy detritus by shallow water flows: an experimental study, *Sedimentary Geol.*, 25, 43-46.
Moss, A.J., Walker, P.H. and Hutka, J., 1980b. Movement of loose, sandy detritus by shallow water flows: an experimental study, *Sedimentary Geol.*, 20, 81-139.
Netto, A.L.C., 1983. Environmental factors affecting discontinuous erosion on forested soils: experimental station of Cachoeira River, Rio de Janeiro, Paper to symposium on The role of geomorphological field experiments in land and water management, Bucharest.
Othieno, C.O., 1979. An assessment of soil erosion on a field of tea under different soil management procedures, *East African Agric. Forestry J.*, 43, 121-127.
Parsons, D.A., 1949. Depths of overland flow, *Soil Conservation Service Technical Paper*, 82, 33 pp.
Pearce, A.J., 1976. Magnitude and frequency of Hortonian overland flow erosion, *J. Geol.*, 84, 65-80.
Pereira, H.C. et al., 1962. Hydrological effects of changes in land use in some East African catchment areas, *East African Agric. Forestry J.*, 27 (Special Issue).
Pinczes, Z., 1982. Variation in runoff and erosion under various methods of protection. *Recent developments in the explanation and prediction of erosion and sediment yield*, IAHS Publication 137, Exeter, 49-57.
Pla Sentis, I., 1980. Soil characteristics and erosion risk assessment of some agricultural soils in Venezuela. In M. De Boodt and D. Gabriels (eds.), *Assessment of soil erosion*, Wiley-Interscience, Chichester, 123-138.
Platford, G.G., 1979. Research into soil and water losses from sugarcane fields, *Proc. South African sugar technologists assoc.*, 1-6.
Platford, G.G., 1982. The determination of some soil erodibility factors using a rainfall simulator, *Proc. South African sugar technologists assoc.*, 1-4.

Poesen, J., 1981. Rainwash experiments on the erodibility of loose sediments, *Earth Surface Processes and Landforms*, 6, 285-307.

Poesen, J., 1983. Influence of slope angle on infiltration rate and Hortonian overland flow volume, *Third Benelux colloquium on geomorphic processes*, University of Liege.

Poesen, J., 1985. An improved splash transport model, *Z. Geomorph.*, 29, 193-211.

Poesen, J., 1986. Field measurements of splash erosion to validate a splash transport model, *Z. Geomorph. Supp. Bd.*, 58, 81-91.

Poesen, J. and Savat, J., 1980. Particle-size separation during erosion by splash and runoff. In D. Gabriels and M. De Boodt (eds.), *Assessment of soil erosion*, Wiley-Interscience, Chichester, 427-437.

Ponce, S.L., 1975. Examination of a non-point source loading function for the Mancos Shale wildlands of the Price River Basin, Utah, Unpublished Ph.D. thesis, Utah State University.

Ponce, S.L. and Hawkins, R.H., 1978. Salt pickup by overland flow in the Price River basin, Utah, *Water Resources Bull.*, 14, 1187-1200.

Quansah, C., 1981. The effect of soil type, slope, rain intensity and their interactions on splash detachment and transport, *J. Soil Science*, 32, 215-224.

Quansah, C., 1982. Laboratory experimentation for the statistical derivation of equations for soil erosion modelling and soil conservation design, Unpublished Ph.D. thesis, Cranfield Institute of Technology.

Rapp, A., Murray-Rust, D.H., Christiansson, C., and Rerry, L., 1972. Soil erosion and sedimentation in four catchments near Dodoma, Tanzania, *Geografiska Annaler*, 54, 255-318.

Reeve, I.J., 1982. A splash transport model and its application to geomorphic measurement, *Z. Geomorph.*, 26, 55-71.

Rensburg, H.J. van, 1955. Runoff and soil erosion rates, Mpwapwa, Central Tanganyika, *East African Agric. J.*, 23, 228-231.

Richter, G., 1978. Soil erosion in Central Europe, *Pedologie*, 28, 145-160.

Richter, G., 1980. Three years of plot measurements in vineyards of the Moselle Region – some preliminary results, *Z. Geomorph. Supp. Bd.*, 35, 81-91.

Richter, G., 1982. Boden erosion in den verschiedenem Boden des Mesozoikums der Trier-Bitburger Mulde in Vergleich zu den Boden des Schiefergebirges. Unpublished manuscript, University of Trier.

Richter, G. and Negendank, J.F.W., 1977. Soil erosion and their measurement in the German area of the Moselle River, *Earth Surface Processes*, 2, 261-278.

Roose, E.J., 1976. Contribution à l'étude de l'influence de la mesofaune sur pédogenese actuelle en milieu tropical, *Rapp. ORSTOM*, Abidjan, 2-23.

Roose, E.J., 1981. Dynamique actuelle de sols ferrallitiques et ferrugineux tropicaux d'Afrique Occidentale, *Travaux et Documents de l'ORSTOM*, 130.

Rowntree, K.M., 1982. Sediment yields from a laboratory catchment and their relationship to rilling and surface armouring, *Earth Surface Processes and Landforms*, 7, 153-170.

Savat, J., 1976. Discharge velocities and total erosion of a calcareous loess: a comparison between pluvial and terminal runoff. *Revue de Géomorphologie Dynamique*, 24, 113-122.

Savat, J., 1977. The hydraulics of sheet flow on a smooth surface and the effect of simulated rainfall, *Earth Surface Processes*, 2, 125-140.

Savat, J., 1979. Laboratory experiments or erosion and deposition of loess by laminar sheetflow and turbulent rillflow, *Seminar on agricultural soil erosion in temperate non-mediterranean climate*, Universite Louis Pasteur, Strasbourg, 139-143.

Savat, J., 1980. Resistance to flow in rough supercritical sheetflow, *Earth Surface Processes*, 5, 103-122.

Savat, J. and De Ploey, J., 1982. Sheetwash and rill development by surface flow. In R.B. Bryan and A. Yair (eds.), *Badland geomorphology and piping*, Geo Books, Norwich, 113-126.

Savat, J. and Poesen, J., 1977. Splash and discontinuous runoff as creators of fine sandy lag deposits with Kalahari sands, *Catena*, 4, 321-332.

Shen, H.W. and Li, R-M., 1973. Rainfall effect on sheetflow over smooth surface, *J. Hydraulics Div. ASCE*, 99, HY5, 771-791.

Schmidt, R.G., 1979. Probleme der erfassung und quantifizierung van ausmass und prozessen der aktuellen boden erosion (abspulung) auf ackerflachen, *Physiogeographica*, Bd. 1.

Schmidt, R.G., 1980. Probleme der simulation erosiver niederschlage, *Regio Basiliensis*, 21, 174-185.

Schmidt, R.G., 1982. Bodenerosionversuche unter kunstlicher Beregnung, *Z. Geomorph. Supp. Bd.*, 43, 67-79.

Schumm, S.A., 1956a. Evolution of drainage systems and slopes in badlands at Perth Amboy, New Jersey, *Bull Geol. Soc. Amer.*, 67, 597-646.

Schumm, S.A., 1956b. The role of creep and rainwash on the retreat of badland slopes, *Amer. J. Science*, 254, 693-706.

Schumm, S.A. and Lusby, G.G., 1963. Seasonal variations in infiltration capacity and runoff on hillslopes of western Colorado, *J. Geophysical Research*, 68, 3655-3666.

Scoging, H., 1980. The relevance of time and space in modelling potential sheet erosion from semi-arid fields. In M. De Boodt and D. Gabriels (eds.), *Assessment of soil erosion*, Wiley-Interscience, Chichester, 349-366.

Scoging, H., 1982. Spatial variations in infiltration, runoff and erosion on hillslopes in semi-arid Spain. In R.B. Bryan and A. Yair (eds.), *Badland geomorphology and piping*, Geo Books, Norwich, 89-112.

Scoging, H. and Thornes, J.B., 1980. Infiltration characteristics in a semi-arid environment, *Proc. symposium on the hydrology of areas of low precipitation*, Canberra, IASH Publication, 128, 159-168.

Slupik, J., 1981. Role of slope in generation of runoff in Flysch Carpathians, *Polish Academy of Sciences, Geographical Studies*, 142.

Soyer, J., Niti, T. and Aloni, K., 1982. Effets et comparés de l'érosion pluviale en milieu péri-urbain de région tropicale (Lubumbashi, Shaba, Zaire), *Revue de Géomorphologie Dynamique*, 31, 71-80.

Staples, R.R., 1934. Runoff and soil erosion tests, *Department of veterinary science and animal husbandry annual reports*, Dar-es-Salaam.

Streumann, Ch. and Richter, G., 1966. Bibliographie zur Bodenerosion in Mittleleuropa, *Berichte Landeskunde Sonderh.* 9, Bad Godesburg.

Sutherland, R.A., 1983. Mechanical and chemical denudation in a semi-arid badland environment, Dinosaur World Heritage Park, Alberta, Canada, Unpublished M.Sc. thesis, University of Toronto.

Swanson, N.P., Dedrick, A.R. and Weakly, H.E., 1965. Soil particles and aggregates transported in runoff from simulated rainfall, *Amer. Soc. Agric. Engineers*, 8, 437-440.

Tefere, F., 1984. The effect of narrow grass strips in controlling soil erosion and runoff on sloping land, Unpublished M.Sc. thesis, Faculty of Agricultural Engineering, University of Nairobi.

Temple, P.H., 1972. Measurements of plot scale with particular reference to Tanzania, *Geografiska Annaler*, 54, 703-220.

Temple, P.H. and Murray-Rust, D.H., 1972. Sheetwash measurements on erosion plots at Mfumbwe, Eastern Uluguru Mountains, Tanzania, *Geografiska Annaler*, 54, 195-202.

Torri, D. and E Sfalanga, M., 1980. Estimate of soil erodibility by simulated rain in laboratory: II Preparation of soil sample, *Instituto sperimentale per la studio a la difesa del Suolo-Firenze* Report.

Ulsaker, L.G., 1982. The design and installation of runoff plot equipment for the National Dryland Farming Research Station, Katumani, Machakos. In B.D. Thomas and W.M. Senga (eds.), *Soil and water conservation in Kenya*, Institute for Development Studies and Faculty of Agriculture, University of Nairobi, 50-64.

Valentin, C., 1979. Problèmes méthodologiques de la simulation de pluies, *Sem. agric. soil erosion in temperate non-mediterranean climate*, Strasbourg-Colmar, 117-122.

Valentin, C., 1981. Organisations pelliculaires superficielles de quelques sols de région subdésertique, Thèse doctorat, Université de Paris VII.

Valentin, C. and Roose, E.J., 1981. Soil and water conservation problems in pineapple plantations of south Ivory Coast. In R.P.C. Morgan (ed.), *Soil conservation: problems and prospects*, Wiley-Interscience.

Van Asch, T.W.J., 1980. Water erosion on slopes and landsliding in a Mediterranean region, *Utrechtse Geografische Studies*, 20.

Van der Linden, P., 1983. Soil erosion in Central Java (Indonesia). A comparative study of erosion rates obtained by erosion plots and catchment discharges. *Catena Supp.*, 4, 141-160.

Wischmeier, W.H. and Smith, D.D., 1965. Predicting Rainfall-Erosion Losses from Cropland East of the Rocky Mountains, *US department agric. handbook*, 282.

Wischmeier, W.H. and Smith, D.D., 1978. Predicting Rainfall Erosion Losses, *US department agric. handbook*, 537.

Wolman, M.G. and Miller, J.P., 1960. Magnitude and frequency of forces in geomorphic processes, *J. Geol.*, 68, 54-74.

Yair, A., 1972. Observations sur les effets d'un ruisellement dirige selon la pente interfluves dans une région semi-aride d'Israël, *Rev. Géog. Physique et Géomorph. Dynamique*, 14, 537-548.

Yair, A., 1973. Theoretical considerations on the evolution of convex hillslopes, *Z. Geomorph.*, 18, 1-9.

Yair, A., 1974. Sources of runoff and sediment supplied by the slopes of a first-order drainage basin in an arid environment, Geomorphologische Prozesse und Prozesskombinationen in des Gegenwent unter Verschiedenen Klimabedungen, (ed. H. Poser), *Abh. der akademie der wissenschaften in Göttingen*, Math-Phys. Kl.III, 29, 403-417.

Yair, A., 1981. The Sede Boker experiment site. In J. Dan, R. Gerson, H. Koyumdjisky and D.H. Yaalon (eds.), *Aridic soils of Israel*, Agricultural Research Organization Special Publication 190, 239-253.

Yair, A., 1983. Hillslope hydrology, water harvesting and areal distribution of some ancient agricultural systems in the northern Negev desert, *J. of Arid Environments*, 6, 283-301.

Yair, A., Bryan, R.B., Lavee, H. and Adar, E., 1980. Runoff and erosion processes and rates in the Zin Valley badlands, Northern Negev, Israel, *Earth Surface Processes*, 5, 205-226.

Yair, A. and Daniu, A., 1980. Spatial variations in vegetation as related to the soil moisture regime over an arid limestone hillside, northern Negev, Israel, *Oecologia*, 47, 83-88.

Yair, A. and Klein, M., 1973. The influence of surface properties on flow and erosion processes on debris covered slopes – an arid area, *Catena*, 1, 1-18.

Yair, A. and Lavee, H., 1976a. Runoff generation process and runoff yield from arid talus mantled slopes. *Earth Surface Processes*, 1, 235-247.

Yair, A. and Lavee, H., 1976b. Trends of sediment removal from arid scree slopes under simulated rainfall experiments, *Hydrol. Sciences Bull.*, 22, 379-391.

Yair, A. and Lavee, H., 1981. An investigation of source areas of sediment and sediment transported by overland flow along arid hillslopes. *Erosion and sediment transport measurement symposium, Florence*, IAHS Publication, 113, 433-446.

Yair, A. and Rutin, J., 1981. Some aspects of the regional variation in the amount of available sediment produced by isopods and porcupines, northern Negev, Israel, *Earth Surface Processes and Landforms*, 6, 221-234.

Yair, A., Sharon, D. and Lavee, H., 1980. Trends in runoff and erosion processes over an arid limestone hillside, northern Negev, Israel, *Hydrol. Sciences Bull.*, 25, 243-255.

Young, A., 1958. Some considerations of slope form and development, regolith and denudational processes, Unpublished Ph.D. thesis, University of Sheffield.

Zachar, D., 1982. *Soil erosion*, Elsevier, Amsterdam.

Zanchi, C., Bassoffi, P., D'Edidio, G., Nistri, L., 1981. Field rainfall simulator. *Erosion and sediment transport measurement symposium*, Field Excursion Guide, IAHS, Florence, 46-49.

CHAPTER 5

Fluvial geomorphology

WILLIAM E. DIETRICH & JOHN D. GALLINATTI
University of California, Berkeley, USA

5.1 INTRODUCTION

Opportunities to do field experiments in order to elucidate the processes controlling river channel form and behaviour are provided by both man-induced and natural circumstances. Dams, channelization, urban water runoff and diversion of flow for irrigation (to name a few) cause changes in the load or discharge characteristics of a reach of river and may generate a useful regularity in flow or channel form similar to that found in the more controlled environment of laboratory flumes. Natural events such as landslides, climate change, tectonic deformation and even volcanic eruptions may induce changes in load, slope or runoff that produce adjustment in channel form or process which in turn shed light on the role of these controlling factors. Fundamental differences in scale, rates and physical interactions between laboratory and field require the field experiment to be the ultimate test of any insight gained from laboratory investigations.

There are problems best addressed by intensive study of a short reach of river such as a bend or braid and there are large-scale issues where long reaches of channel must be investigated. At the field scale one is confronted with several difficult problems. Stream flows are unsteady, nonuniform and typically too deep and murky to observe directly processes near the bed and bank. Yet, it is these boundary interactions that lead to morphologic adjustment of the channel. Although compared with other geomorphic features river channels adjust and evolve rapidly, their size and the unsteady nature of the flow and sediment transport make accurate measurement difficult. Hence, often the most critical decision is the selection of the study site.

In all cases, field experiments in fluvial processes are most successful when they are guided by and influence the development of quantitative theories. These theories may be largely empirical in their constructions in which case correlation is sought between presumed major controls and observed morphology pattern or process. They may also be strongly physically-based, in which case fluid and sediment transport mechanics are used to formulate specific predictive equations. Although the empirical approach, which predominates, often leads to valuable observations it rarely provides unequivocal explanations. The more physically-based approach is slowly providing answers to major problems, although it seems often the case that the apparent answer lies hidden in obscure mathematics rather than in simple physics.

As reflected in several major international conferences in recent years, considerable research is now being done in fluvial geomorphology throughout the world. Recent symposium volumes include: International Association of Hydrologic Sciences (1981a, b and 1982; numbers 132, 133 and 137), Proceedings of Euromech conferences, International Association of Sedimentologists Special Pub. (6) (1983), International Symposium on River Sedimentation, China (1983), Gravel-bed Rivers (Hey et al. 1982) and River Meandering (Elliot 1984). Useful literature reviews are reported in the recent books by Richards (1982) and Allen (1982). In this paper we do not attempt to compile an exhaustive list of all individuals and researchers involved in field experiments in fluvial geomorphology. With a generous definition of 'experiment' the list would be very long and include individuals from quite diverse fields such as civil engineering, physical geography, sedimentology, aquatic biology and geomorphology. Instead we have elected to focus on what appear to be emerging new issues in river channel studies that are attracting international attention. An attempt has been made to focus on research currently being done and only recently reported, and undoubtedly individuals or research groups have been missed, particularly in the non-English publishing countries. To these people we apologize for our oversight. Also, some areas of fluvial geomorphology, such as the study of channel networks, are not examined here. In this latter case, the review by Abrahams (1984) is very useful.

This review reflects our bias that field experiments should be rooted as firmly as possible in an understanding of fluid and sediment transport mechanics. We begin with a discussion of field studies of local flow structure, boundary shear stress distribution and sediment transport. From there we progress to problems concerning bed morphology, planform and behaviour. The approach is entirely process-oriented, in keeping with the notion of a field experiment.

5.2 CHANNEL MECHANICS

The shape of a river channel is a consequence of the interaction of flow with its deformable boundary. Conceptually, this interaction can be separated into three general processes (Fig. 5.1). Flow in a river channel reacts to the channel topography (such as curvature, bank irregularities, bars and pools, dunes, and bed material) and generates a boundary shear stress field (force/unit area acting parallel to the bed surface) which in turn controls the pattern of sediment transport. The channel morphology is adjusted and

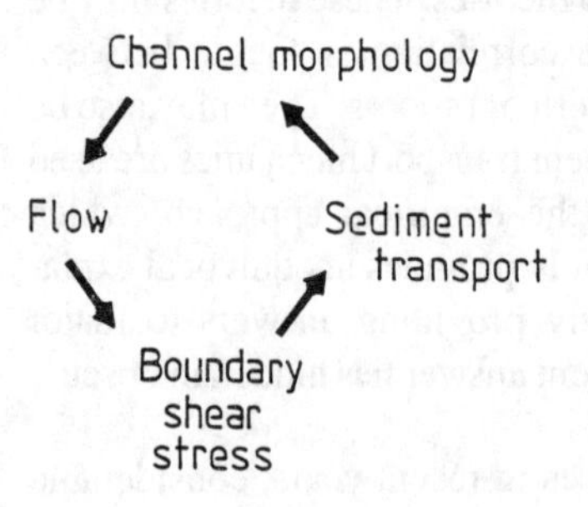

Figure 5.1. Processes controlling river morphology. Flow reacts to channel morphology (including particle size) to generate a boundary shear stress field. The boundary shear stress causes sediment transport. The downstream and cross-stream changes in sediment transport lead to morphologic adjustment. Given sufficient duration of constant flow and sediment supply, an equilibrium morphology may develop through this interaction. Not indicated in this diagram is an important direct interaction between the moving sediment, the flow, and the boundary shear stress.

brought toward equilibrium with the prevailing flow conditions and boundary shear stress field by the downstream and cross-stream patterns of sediment transport. Where the boundary shear stress (in excess of that necessary to cause sediment movement) diminishes in the downstream direction, either net deposition will occur or sediment must be transported across the channel away from the decreasing shear stress zone. Similarly along zones of increasing boundary shear stress, net erosion will occur or there must be compensating cross-stream transport into the zone of increasing boundary shear stress. In both cases, bed material grain size variations may reduce or prevent the requisite net erosion or deposition. Cross-stream sediment transport results from a cross-stream component of flow and boundary shear stress and from the cross-stream component of gravity acting on the bed particles lying on a sloping surface.

In order to understand, quantify and predict the behaviour and morphology of river channels the interaction depicted in Figure 5.1 must be examined theoretically and with measurements. Ideally these are done together so that models are based on correct physical assumptions, and useful measurements are taken with sufficient accuracy and thoroughness to serve as tests of models. The general equations that must ultimately underlie all models are the equations of motion for a general fluid flow:

$$\rho \frac{d\pi}{dt} = -\nabla p + \nabla \tau - \rho g \tag{5.1}$$

the continuity equation of fluid flow and sediment transport

$$\nabla \pi = 0 \tag{5.2a}$$

$$\nabla Q_t = -C_b \frac{\partial \eta}{\partial t} \tag{5.2b}$$

and an equation relating flow conditions to boundary shear stress computed in Equation 5.1 which in turn is related to the local sediment transport rate

$$\tau_b = f\left(\frac{\partial \pi}{\partial z}\right) \tag{5.3a}$$

$$Q_t = f(\tau_b, D, \ldots) \tag{5.3b}$$

Here π is the velocity vector, t and ρ represent time and density respectively, p is pressure, τ is the total deviatoric (non-isotropic) stress tensor, g is the gravitational acceleration, Q_t is the sediment transport vector, C_b is the concentration of sediment in the bed (1-porosity), η is the height of the bed above an arbitrary datum, τ_b is the boundary shear stress vector, z is the height above the boundary, and D is a representative bed particle size.

In the last five years impressive progress has been made in both the modelling and field measurement of the elements in these equations. The discussion to follow will describe recent and ongoing field studies.

5.2.1 *Flow processes and patterns*

Most river channels have bar and pool topography and are not straight. Rapid changes in bed elevation and bank alignment cause large momentum forces and pressure gradients in channel flow that result in cross-stream flow and that substantially influence the magnitude and direction of the boundary shear stress. In order to document these effects detailed velocity vector measurements at sequential, closely spaced cross-sections are required. Few data of this nature have been collected, in part because of the difficulty of instrumentation, but also because the need for highly accurate, thorough data was not clearly recognized. The need for such data is perhaps most explicitly and simply expressed in the vertically averaged force balance equations derived by Smith and McLean (1984) from the equation of motion (Eq. 5.1) in a curvilinear coordinate system

$$(\tau_{zs})_b = -\frac{\rho g h}{(1-N)}\frac{\partial E}{\partial s} - \rho\frac{1}{(1-N)}\frac{\partial}{\partial s}<u_s^2>h$$
$$-\rho\frac{\partial}{\partial n}<u_s u_n>h + 2\rho\frac{<u_s u_n>h}{(1-N)R} \quad (5.4)$$

$$(\tau_{zn})_b = -\rho g h\frac{\partial E}{\partial n} - \rho\frac{<u_s^2>h}{(1-N)R} - \frac{1}{(1-N)}\frac{\partial}{\partial s}<u_s u_n>h$$
$$-\rho\frac{\partial}{\partial n}<u_n^2>h + \rho\frac{<u_n^2>h}{(1-N)R} \quad (5.5)$$

In their coordinate system, s is in the downstream direction and parallel to the channel centerline. The n axis is perpendicular to the s axis in the cross-stream direction, and it is positive toward the left bank. The z axis is nearly vertical. The scale factor in the downstream direction is $1-(n/R) = 1 - N$, where R is the local radius of curvature of the centerline; the other scale factors are unity. In the above equations $(\tau_{zs})_b$ and $(\tau_{zn})_b$ are the downstream and cross-stream components of boundary shear stress, h and E are the depth of flow and elevation of the water surface with respect to an arbitrary datum, and u_s and u_n are the downstream and cross-stream components of the velocity. The bracket, < >, indicates that the enclosed quantity has been vertically averaged. The first term on the right hand side of Equation 5.4 represents the downstream pressure gradient force; the next two express the change in momentum of the downstream flow in the downstream and cross-stream directions, respectively, and the last term represents the force associated with channel curvature. The cross-stream force balance, expressed by Equation 5.5,

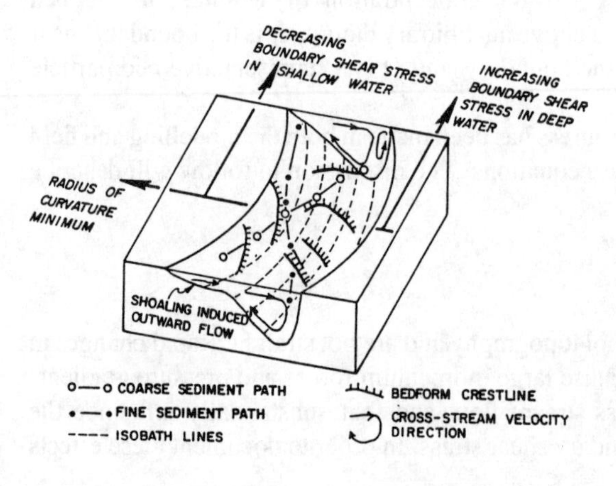

Figure 5.2. Processes controlling the equilibrium bed topography and sorting of sediment in a sand-bedded river bend. Outward sediment transport across the top of the point bar and against the inward near-bed secondary flow leads to a balance between the local sediment supply and downstream increases and decreases of boundary shear stress (from Dietrich and Smith, in press).

indicates that the cross-stream boundary shear stress is equal to the sum of forces caused by the cross-stream pressure gradient, by the centrifugal acceleration of the flow (second and fifth terms on the right hand side) and by the change in momentum in the cross-stream direction (third and fourth terms).

The above equations apply to straight channels as well as curved ones. They should also apply to individual subchannels in braided reaches. The effects of bed topography are explicitly shown in spatial derivative terms involving $\frac{\partial h}{\partial s}$ and $\frac{\partial h}{\partial n}$, the downstream and cross-stream bed slopes. Because of the contribution of $<u_s u_n>$ to the overall force balance and because of the importance of the cross-stream component of flow and boundary shear to the sediment continuity Equation 5.2, the cross-stream component of water velocity should be carefully evaluated.

In order to employ the force balance Equations 5.4 and 5.5 to examine flow and channel topography interactions with field measurements, it is essential that the cross-sections along which the field data are collected be correctly oriented with respect to channel topography (see Dietrich and Smith 1983). This requires that cross-stream discharge data collected at successive cross-sections be compared with that needed to satisfy the continuity equation:

$$<u_n>h = -\frac{1}{1-N}\int_{-w/2}^{n} \frac{\partial <u_s>h}{\partial s}\, dn \qquad (5.6)$$

By integrating Equation 5.6 across the channel from $-w/2$ (where w is the channel width) to $+w/2$, the cross-stream discharge can be computed and compared with observed. The cross-section orientation can then be corrected so that the cross-section discharge equals that required by continuity. Dietrich and Smith (1983, 1984a) discuss the importance of using the continuity equation to examine the flow through a river meander. Its proper use leads to finding that in river meanders with bar and pool topography there is a tendency for shoaling over the upstream part of the bar to cause net outward discharge into the pool. The classical helicoidal cross-stream flow pattern of near surface flow toward the outside concave bank and near bed flow toward the inside bank was confined to the deepest 20 to 30% of the channel width (Fig. 5.2). Dietrich and Smith (1983, 1984a) have proposed that the near-bed outward flow and resulting cross-stream sediment transport are required for the formation of an equilibrium bed topography during periods of significant transport. Thorne and Rais (1984) have subsequently obtained very similar cross-stream flow patterns during low flow of the Fall River in Colorado (Fig. 5.3). Thorne's group at Colorado State University plan further measurements in the Fall River in subsequent years. Siegenthaler (1984) is completing a master's thesis at Colorado State University in which he will review and evaluate presently available methods for measuring velocity patterns in the field. He is also analysing the relative error that can be expected in the calculation of boundary shear stress from field observation using the terms in the Smith-McLean force balance Equations 5.4 and 5.5 and similar equations by Kalkwijk and De Vriend (1980). A summary of findings is reported by Siegenthaler and Shen (1984).

Other measurements of flow through river meanders include Bridge and Jarvis' (1982) observations over a range of stage, in the South Esk River, Scotland, Hooke and Harvey's (1983) data on downstream and cross-stream flow in the gravel bedded River Dane in England, and De Vriend and Geldof's (1983) detailed downstream velocity fields in a

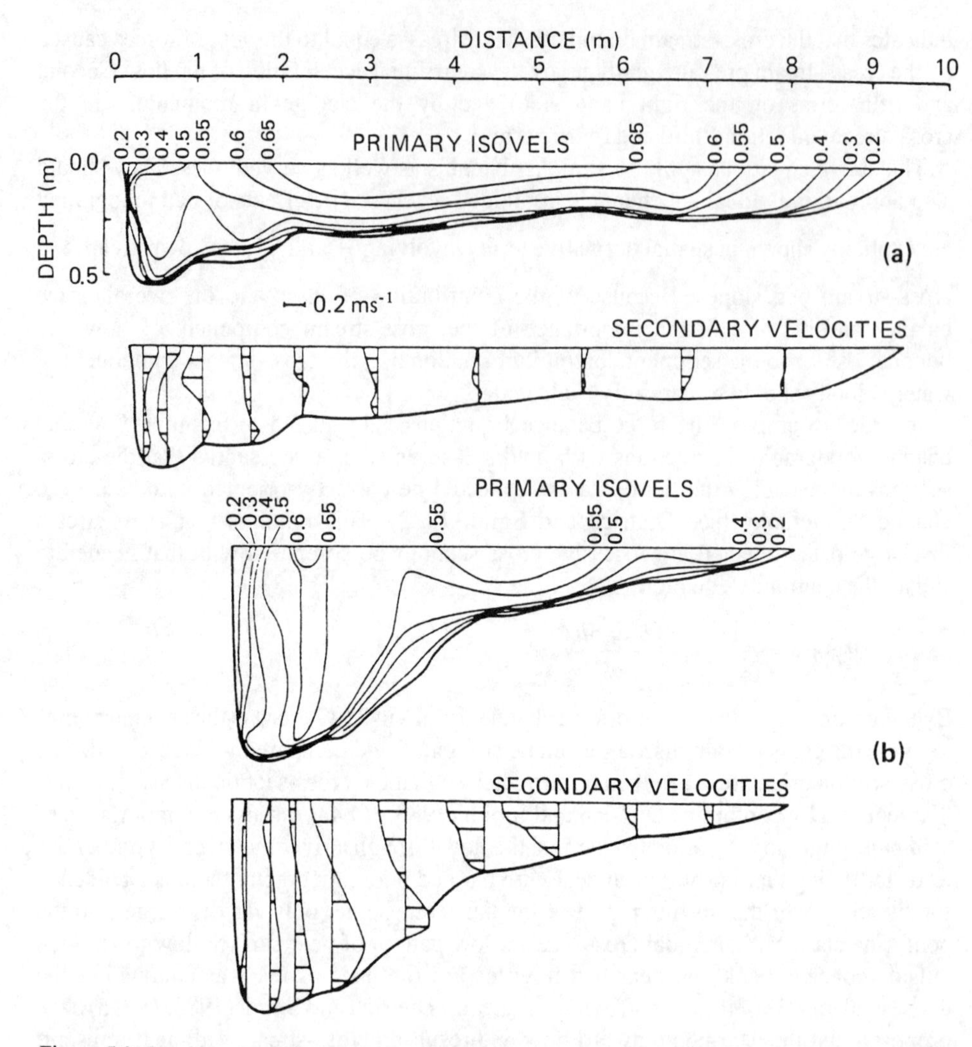

Figure 5.3. Downstream and cross-stream components of velocity at 2 sections on the Fall River, Colorado. (a) Section 1: bend A, entrance. (b) Section 2: bend A, apex. Measurements were made at low stage in a reach recently aggraded with excessive and supplied from a dam failure (Thorne and Rais 1984).

meander of the River Dommel in the southern Netherlands. Chiu et al. (1984) have measured the velocity field in a bend of the East Fork River, Wyoming and compared the observations with predictions from a mathematical model for flow and boundary shear stress.

Smith and McLean (1984) used their mathematical model for flow through curved channels to perform numerical experiments which shed insight into field studies (Fig. 5.4). They systematically varied the amplitude of the bed topography (from flat bed to natural bar-pool topography), the strength of turbulent mixing, the boundary roughness

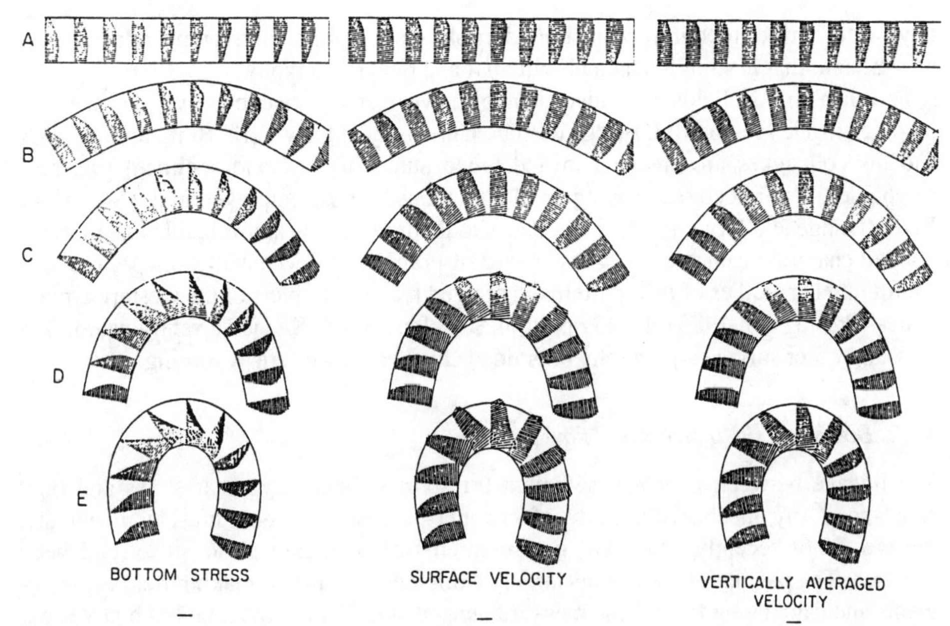

Figure 5.4. Predicted variation of flow properties with sinuosity in a channel with typical bed topography, according to the Smith-McLean model. The values of H_0 (the angular deviation between the down valley axis and the channel path at the geometric crossing between the bends) in this figure are 0° (A), 30° (B), 55° (C), 80° (D) and 105° (E). Note that bed topography effects dominate over those due to channel curvature for < 30° or so; whereas, the curvature becomes quite important for w > 55° (from Smith and McLean 1984).

and the planform sinuosity to examine their relative effects on the boundary shear stress and flow fields. The predominant control on the pattern of boundary shear stress was the amplitude of the point bar in the bend. A well-developed point bar forces the locus of maximum boundary shear stress to shift outward through the bend and causes the shear stress vector over the point bar top to be oriented toward the concave bank. Increased turbulent mixing was found to increase the magnitude of the boundary shear stress and reduce the secondary circulation. Greater roughness resulted in greater values of boundary shear stress but did not change the structure of the shear stress field. Increasing the curvature of the channel with well-developed bar-pool topography had a much smaller effect on the boundary shear stress field than changes in the bed topography. Evidence for most of these predicted changes can be found in a survey of available flume and field data.

Detailed field studies of flow patterns have focused on river meanders almost to the exclusion of straight and braided channels. An exception is the study by Leopold (1982) of nearly straight sections of Baldwin Creek and studies of straight sections with bar and pool topography would furnish insight into the influence bed topography has in the absence of channel curvature on flow patterns and boundary shear stress fields. Leopold found patterns of water surface topography and flow comparable to that found in meandering reaches. Theoretical analysis by Smith and McLean (1984) indicates that flow, water surface topography and boundary shear stress found in river meanders should be similar to that in straight channels with bar and pool topography.

Few data on flow fields in braided channels have been reported, but several groups are involved in such research. J. Bridge of State University of New York, Binghamton, USA and his colleagues are engaged in a detailed study of flow and sediment transport mechanics in a braided reach using similar methods to those employed in the South Esk River (Bridge and Jarvis 1982). Their data should provide new and valuable insights into braided channel mechanics. K. Prestegaard of Franklin and Marshall College, USA, is continuing her studies of flow patterns in braided reaches in rivers of the western United States (Prestegaard 1983a, b). Leopold of U.C. Berkeley, USA, is currently examining flow and water surface topography in braided channels of western Wyoming.

5.2.2 *Boundary shear stress and flow resistance*

The linkage between flow and sediment transport is boundary shear stress and flow resistance. Any mechanistic study of channel process and form must confront this linkage. Until recently, few field measurements of boundary shear stress had been reported. The paucity of data stems largely from the difficulty of making field measurements and calculating from these measurements that portion of the total boundary shear stress responsible for sediment transport. Although the total boundary shear stress is an important element in understanding river channel processes, it is that component of the boundary shear stress that causes sediment transport that needs to be documented and related to channel morphology. Here we review four methods that have been used recently to map boundary shear stress in river channels. Several useful discussions of field methods and data interpretations regarding boundary shear stress and flow resistance in gravel-bedded rivers can be found in the book edited by Hey et al. (1982).

In steady, uniform flows in straight channels with flat beds the boundary shear stress equals the downslope component of the weight of the fluid:

$$\tau_b = \rho g h S \tag{5.7}$$

where S equals the water surface slope. If two dimensional bedforms, such as ripples or sandwaves are present, the total boundary shear stress can still be calculated from Equation 5.7, but a substantial portion of the total stress may arise from form drag over the bed features and not be applied to the transport of bed particles. Several methods have been proposed to compute the form drag component of the boundary shear stress (see review in Raudkivi 1976, Engelund and Fredsoe 1982, and Brownlie 1983), but few tests of the models have been performed in natural rivers. Bridge and Jarvis (1982) have attempted to apply the Engelund and Fredsoe (1982) model for bedform resistance to observations on the South Esk River, Scotland. Dietrich et al. (1984) have found that the model proposed by Smith and McLean (1977) and employed on the Columbia River predicts fairly accurately the form-drag-induced boundary shear stress over dunes in Muddy Creek, Wyoming, a small sand-bedded river. The ratio of boundary shear stress associated with bedform drag and grain resistance to that just due to grain (moving and stationary) resistance was computed from field data to be 1.8-2.0 (low flow to high flow), the model predicted 2.2-2.3. More work in sand-bedded streams is needed to test further the utility of the Smith-McLean and other models.

Most rivers have beds that are deformed into three-dimensional large-scale bedforms, or bars. As discussed below, these bars typically possess a scour hole lobe-front morphology: the holes appearing on alternate sides of the channel in the downstream direction. Although most clearly developed in gravel-bedded rivers, this larger scale bed form is characteristic of sand-bedded rivers as well (see Fig. 13.1, Leopold 1982). Bar and pool topography greatly affects the pattern of boundary shear stress and as for the smaller-scale bedforms, the form drag associated with flow over these bed features must be computed and subtracted from the total boundary shear stress.

Prestegaard (1983a) has found a correlation between the component of shear stress associated with bar-pool topography and the ratio of the elevation difference between the pool maximum depth and top of bar and the distance between successive bars. Lisle (1982) observed that aggradation of channels in northern California following a major flood led to reduced bar amplitude. He proposed that this would in turn reduce the form drag resistance due to bar-pool topography and enhance bedload transport rates. Dietrich et al. (1984) calculated that the form drag over the bar-pool topography was equal to nearly half the total boundary shear stress at high flow and only about 20% at low flow in a sand-bedded river meander. These values were reasonably well predicted with the Smith-McLean equation.

Recently, maps of the water surface topography have been used in conjunction with depth of flow measurements to compute from Equation 5.7 the spatial pattern of total boundary shear stress through bends (Dietrich et al. 1979, 1984, Bridge and Jarvis 1982, and Leopold 1982). Dietrich et al. (1979, 1984) have shown that there is only a crude correlation, however, between the pattern of boundary shear stress estimated from Equation 5.7 and that computed from other more direct means. Large forces arising from spatial acceleration of the flow over the bar and pool topography directly influence the calculation of boundary shear stress from the measured water surface topography. Application of Equation 5.4 to detailed field measurements in Muddy Creek, Wyoming demonstrates that the convective acceleration terms associated with changing bedslope are equal to or greater than the downstream pressure gradient force term. Although the two spatial acceleration terms tended to be of opposite sign and cancel each other, this was not always the case. Unfortunately, accurate calculation of the convective acceleration term in Equation 5.4 appears to require a density of fluid measurement that is difficult to obtain in practice. Studies of channel mechanics require more direct measures of boundary shear stress.

In recent field studies renewed interest has developed in the use of the 'law of the wall' equation to compute local boundary shear stress (Bathurst et al. 1979, Bridge and Jarvis 1982, Dietrich et al. 1979, 1984, Rohrer 1984). The equation can be written as

$$u = \frac{u_*}{k} \ln z/z_0 \tag{5.8}$$

where u is the velocity at height z above the bed, u_* is the shear velocity, equal to $(\tau_b/\rho)^{1/2}$ is the local boundary shear stress, k is von Karman's constant and z_0 is a roughness parameter. Improved current meter technology makes application of this equation promising for field studies concerned with mechanics.

Several problems, however, can arise. Velocity measurements should not be used in Equation 5.8 for heights greater than about 0.2 the depth because in this interior region the assumptions used to derive Equation 5.8 break down. The current meter must be highly accurate and yield data with high precision. In many flows, errors much greater than ± 1.0 cm/sec may severely reduce the accuracy of the computed boundary shear stress. Similar requirements must be met for locating the current meter in the flow. Little work has been done on gravel-bedded rivers, but it is the experience of Luna Leopold (U.C. Berkeley) that long sampling periods (about five minutes) may need to be used at each height above the bed if a single current meter is used to document the local velocity profile. When particle sizes approach the total flow depth, applicability of Equation 5.8 diminishes (see Bathurst et al. 1981 for a discussion of this case).

Traditionally it has been argued that von Karman's constant, which is 0.4 in clear water flow, reduces to values below 0.2 due to stratification effects of suspended sediment (see review in Vanoni 1976). Recently, several independent theoretical studies have shown that von Karman's constant does not vary (see review in paper by Bridge and Dominic 1984). Instead it appears that the velocity gradient in the region of large vertical decline in suspended sediment concentration is increased due to reduced turbulent exchange caused by density stratification. This means that velocity profiles taken under conditions of high suspended sediment loads will not yield estimate of boundary shear stress correctly from Equation 5.8 alone. Corrections must be used, according to the concentration gradients. This problem is discussed further by Smith and McLean (1977b) and Nikitin and Deineka of the USSR (1984).

Finally, work by Gust and Southard (1983) suggests that even in the case of purely bedload transport where the particles are in low concentration and are rolling on the bed, significant deviations from the clear water logarithmic profile develop that may not be attributable to increased resistance to flow caused by the moving grains. For discussion of the effects of moving grains on the roughness parameter see review by Bridge and Dominic (1984) and also paper by the Polish scientist Gladki (1979).

Despite these numerous cautionary statements, when used carefully the logarithmic Equation 5.8 provides the most reliable, easily used method to compute boundary shear stress.

If it is assumed that Equation 5.8 applies throughout the flow depth and that the flow is hydraulically rough, then the vertically averaged velocity, $\bar{u}$, occurs at $z \approx 0.368h$, $z_0 \approx D_x/30$ and Equation 5.8 becomes

$$\frac{\bar{u}}{u_*} = \frac{1}{k} \ln \frac{h}{AD_x} \tag{5.9}$$

where D_x is a representative particle size on the bed. The constant, A, is empirically derived and depends on several factors. If in Equation 5.9 $\bar{u}$, and h apply only to a local vertical average at some point in a stream then A and D_x represent local particle size and bedform (dunes and bars) geometry. Dr. Becchi of the University of Florence has recently proposed (1983) that loose, natural beds have irregularly distributed particle groupings that increase significantly the resistance above that of densely packed, flat-topped bed of grains. If $\bar{u}$ and h are the cross-sectional average values and in this case A and D_x include all scales of boundary resistance. It is obviously crude to characterize the bed particle size for a reach channel with a single grain diameter, however, it is impractical to include the spatial variation of size. Bray (1980, 1982) discusses this problem further. Some (Parker and Peterson 1980, Prestegaard 1983a) have proposed that Equation 5.9 be used at high flow as a measure of grain resistance and as a way to compute boundary shear stress responsible for sediment transport. But in Equation 5.9 when $\bar{u}$ and h are cross-sectional averages, other length scales of resistance influence the values of A and D_x and reduce

Equation 5.9's utility as means to compute local boundary shear stress. It would greatly facilitate studies of gravel-bedded rivers if a form of Equation 5.9 could be used with a single, near-bed velocity measurement such that rapid, relatively accurate estimates of boundary shear stress could be made.

A direct measurement of the boundary shear stress can be made by using coupled current meters with high frequency response to record the mean and fluctuating components of velocity. The major components of shear stress in open channel turbulent flow arise from the vertical flux of horizontal momentum and can be computed from velocity measurements and

$$\tau = \rho u'w' \tag{5.10}$$

where u' and w' are the root mean square of the velocity fluctuations about the mean of the downstream and vertical velocity components respectively. McLean and Smith (1979) have used five sets of three orthogonally mounted current meters attached to a 2.5 m frame in the Columbia River to study the turbulent structure of flow over large-scale sand waves. With relatively large (4 cm) mechanical current meters they were able to obtain unique field data documenting the vertical increase in shear stress above the bed due to form drag over the bedforms. Continued advances in current meter and data recording instrument technology should provide the opportunity to make comparable measurements in smaller rivers. This could prove to be the most reliable method to determine boundary shear stress in rivers.

5.2.3 *Sediment transport*

Most models that predict river channel morphology and behaviour require very specific patterns of the direction and rate of sediment transport. Yet, there are virtually no field data on sediment transport fields for reaches of rivers. It is likely that in the near-future this major deficiency will be reduced as methods for making bedload measurements become more firmly established.

Current research on sediment transport processes in river channels can be divided into three broad categories: initial motion studies, local sediment transport measurements, and determination of sediment loads.

5.2.4 *Initial motion*

Natural rivers are typically poorly sorted and the initial motion of bed particles appears to be influenced by packing, shape and relative size of individual particles to surrounding ones (see paper by White and Day 1982, and the discussion papers that follow it). The Shields diagram, which relates the dimensionless critical shear stress $\tau_{*c} = \tau_c/(\rho_S - \rho)gD$ to a bed grain-diameter and specific gravity for a fluid of a given kinematic viscosity and shear velocity, has undergone considerable re-evaluation in recent years (see brief reviews in Shen and Lu 1983, Mantz 1983, and Bridge and Dominic 1984). Some authors have argued that the empirical evidence suggests a constant τ_{*c} of about 0.03 for hydraulically transitional and rough flows. Theoretical analysis, however, argues strongly for τ_{*c} being higher in hydraulically rough flow than in transitional flow. When bed materials are of uniform grain size and with a density of about 2.65 gm/cc, sand size particles lie in the transitional flow region where there is little disagreement. Values of τ_{*c}

for gravel (with heterogeneous size distribution) range from 0.047 (Yalin and Kanahan, 1979) to .06 (Vanoni 1964, Cavazza 1981). All studies indicate a progressive increase in τ_{*c} with diminishing particle size below sand size (Mantz 1983). Stelczer (1981) reviews several basic issues in bedload transport including initial motion and provides valuable discussion of European and Russian studies. The definitive study of this problem is still needed. Attempts to add data collected in the field to the hydraulically rough end of the graph must be viewed with caution. Normally it is assumed that Equation 5.7 provides an accurate estimate of the boundary shear stress responsible for bedload transport, but as we have already discussed, in most natural channels this estimate is crude.

A growing number of authors (e.g. White and Day 1982, Parker et al. 1982, Gomez, 1983, Carling 1983, Andrews 1983, 1984, David and Gangadharaiah 1983, Misri et al. 1983) are addressing the difficult problem of initial motion in poorly sorted sediment. Armouring or paving of the surface is typical in such material. Although there seems to be general agreement that the more exposed coarser particles move at lower shear stres ses than the uniform grain size case (see Fenton and Abbott 1977) and the finer fraction requires higher shear stresses, there is not yet a consensus on how to express this effect quantitatively for nonuniform size sediment.

The laboratory studies and field data analysis of Parker et al. (1982) and Parker and Klingeman (1982) shed new light on the role of pavement in poorly sorted, gravel-bedded streams. They observed that in a flume with a constant sediment feed and water discharge, a paved equilibrium bed surface developed in which the bedload and the sub-pavement sediment had similar grain size distributions. The pavement, which had over twice the median grain size of either the bedload or sub-pavement, clearly exchanged grains with the bedload. To explain this observation Parker and Klingeman (1982) proposed that the formation of a paved surface increases the mobility of the coarse particles and reduces that of fine particles relative to the ambient boundary shear stress such that the poorly sorted sediment supplied to the channel is transported through a reach without net deposition. The inherent mobility difference between coarse and fine particles in a bed of constant grain size is compensated at the bed surface in poorly sorted sediment by increased exposure and concentration of the coarse particles and reduced exposure and concentration of fine particles. Thus, Parker and his colleagues view the formation of a paved surface as an adjustment like that of width or depth changes which drives rivers towards equilibrium form. Analysis of the thorough data set on bedload transport collected by Milhous (1973) on Oak Creek supported their hypothesis of similar mobility for coarse and fine sediment and allowed them to construct a set of empirical expressions to predict bedload and pavement size distributions in Oak Creek.

Following an approach similar to Parker and Klingeman (1982), Andrews (1983, 1984) has used field observations to define an empirical expression relating the critical Shields number of a particular particle size (i), τ_{*ci}, to the ratio of that particle size D_i and the median particle size of the sub-pavement, D_{50}:

$$\tau_{*ci} = \frac{\tau_{ci}}{(\rho_s - \rho)gD_i} = 0.0834\left(\frac{D_i}{D_{50}}\right)^{-.872} \tag{5.11}$$

His equation is quite similar to others relating τ_{*ci} to the ratio of the individual particle size and arithmetic mean size of the surface material. Figure 5.5 is modified from the paper by Misri et al. (1983) and shows the similarity of four different published expressions, particularly for the case of D_i/D_a (average bed particle size) or D_i/D_{50} less

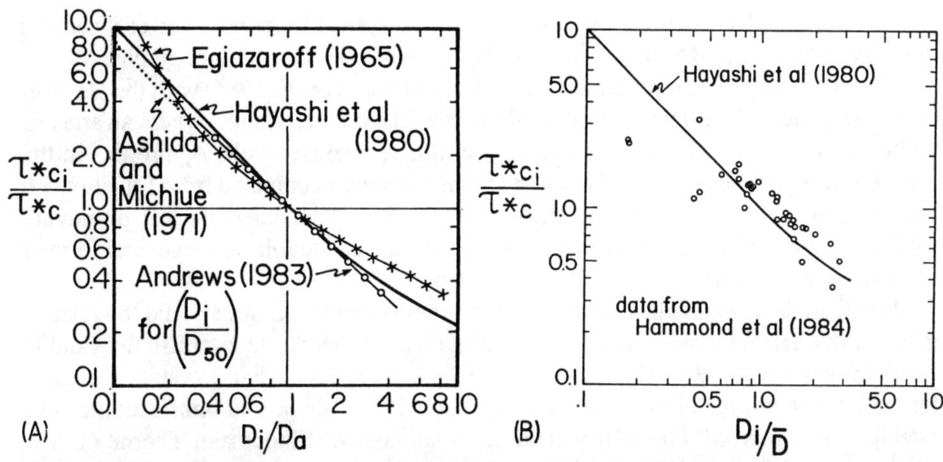

Figure 5.5. Relationship between critical Shields number ratio and bed particle size ratio for heterogeneous particle size distribution. (A) Comparison of four proposed curves. The first three use the ratio of the individual particle size (D_i) to the average size on the bed surface (D_a) and τ_{*c} in this case is the critical Shields number for D_a. The curves are taken from Figure 2 of Misri et al. (1983). Andrews (1983) uses the ratio of the individual particle size to the median in the bed just below the paved surface, D_{50}, and in this case τ_{*c} is for D_{50}. (B) Comparison of Hammond et al. (1984) data on initial motion in a gravel-bedded tidal channel and the Hayashi et al. (1980) curve. The average particle size of the bed material, $\bar{D}$, was 1.7 cm.

than 1.0. Misri et al. report flume data that closely follow the Hayashi et al. (1980) curve as long as τ_{*ci} is allowed to vary empirically with D_a.τ_{*ci} in the Misri et al. experiments ranged from 0.023 to 0.03, Ashida and Michiue (1971) used 0.040 and Andrews (1983, 1984) used 0.0834. Because $\tau_{*ci} = \tau_{*c}$ when $D_i = D_{50}$ (or $D_i = D_a$ in the other studies) then Equation 5.11 simplifies to

$$\frac{\tau_{ci}}{\tau_{c50}} = \left(\frac{D_i}{D_{50}}\right)^{0.128} \tag{5.12}$$

where τ_{ci} and τ_{c50} are the dimensional critical boundary shear stress for the individual particle size and the subpavement median respectively. Over the range of proposed applications of Equation 5.11 ($0.3 < D_i/D_{50} < 4.2$) the ratio in Equation 5.11 varies only from .85 to 1.22, that is particles have nearly the same critical boundary shear stress over a broad range of sizes in a heterogeneous bed. A similar conclusion would be drawn from analysis of the other expressions plotted in Figure 5.5.

Although potentially very useful some important problems remain before Figure 5.5A or Equation 5.11 can be put to general use. First there is the problem of whether to use D_a, D_{50} or some other reference particle size. The median particle size of the sub-pavement, D_{50} is certainly the least practical to measure, and because these are empirical expressions (due to the need to set the value of τ_{*c}), there seems to be little justification for using D_{50}. If τ_{*c}, D_a, and τ_{ci} are held constant then data plotted in Figure 5.5A should tend to produce a slope close to −1.0 because D_i is in the denominator in τ_{*ci} and the numerator in D_i/D_a. Care must be taken to avoid spurious correlations with the coordinates used in

Figure 5.5A. Finally, it is often the case that τ_{ci} is computed from ρghS and as discussed this may be a poor approximation in natural rivers with bar and pool topography.

In Figure 5.5B we compare data reported by Hammond et al. (1984) with the Hayashi et al. (1980) curve. The data are from a 20 m deep tidal marine channel where an array of 6 Ott current meters was used to obtain one minute-averaged velocity profiles in the logarithmic region near the bed. Boundary shear stress was computed from Equation 5.6 and a video camera placed near the gravel bed was used to document the size of moving particles. Although the scatter is considerable, the data tend to follow the general form of the Hayashi et al. curve.

More field data, where the boundary shear stress is known accurately and the relative mobility of different grain sizes is directly observed, are needed to establish the validity of this important approach. Hopefully, a more direct physical linkage accounting directly for the relative protrusion (i.e. Fenton and Abbott 1977), shape and imbrication can be established. It is worth noting that field observations by Milhous and Thorne (1982), Klingeman and Emmett (1982) and Hey (1982) suggest that there is differential mobility of bed particles over the armour layer. Whether this can be accounted for by equations of the form in Equation 5.11 is not yet explored.

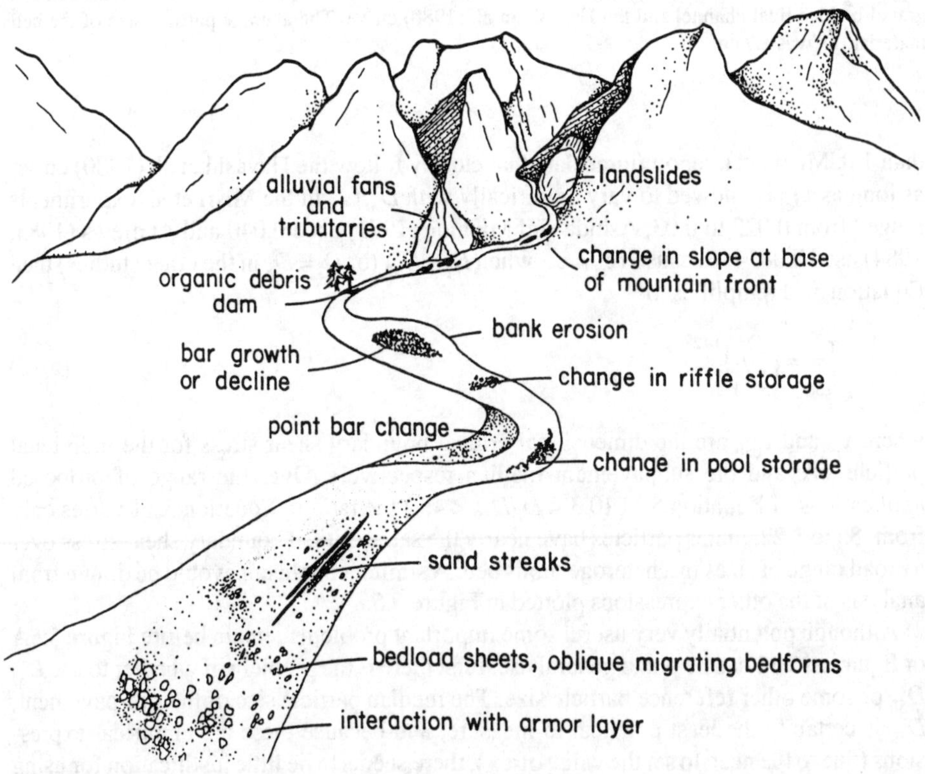

Figure 5.6. Processes that contribute to long and short term variation in sediment supply in rivers. Measurements of sediment transport, particularly bedload, may be strongly affected by upstream changes in input, storage and local bed interactions. Ideally such sources of variation should be documented as part of the measurement program.

5.2.5 *Local transport measurements*

In either laboratory or field studies, bedload transport is found to be quite variable, even when apparent major controls such as boundary shear stress, discharge or sediment supply are kept reasonably constant. For example, in a laboratory study of gravel transport, White and Day (1982) found that the coefficient of variation in measured bedload transport during constant conditions was about 20 to 25%. It can only be expected that under less controlled conditions found in field studies the variance will be as high or higher. Dietrich and Smith (1984) collected sevenhundred and forty bedload samples and made nearly threehundred separate bedform migration measurements at nine sections across a small sand-bedded channel during a period of essentially constant discharge. They found the co-efficient of variation in the computed average cross-sectional bedload transport to be 25%, whereas it was 16% for the bedform measurements. Because the discharge was artificially controlled, these data probably represent the minimum variation that can be obtained with this sampling procedure. Under more natural flow, variation for intensively sampled channels probably is not easily reduced below 50%.

Increasingly, field studies are pointing to the importance of periodic sediment supply on observed transport rates and channel behaviour (Fig. 5.6). Church and Jones (1982) and Church (1982) have argued that there is a tendency for 'sedimentation zones' to develop at regular intervals along gravel bed rivers. These zones, characterized by apparent channel instability, braiding and abundant movable sediment, may be initiated by a 'slug of sedimentary debris' introduced rapidly into the channel from tributaries, landslides, or other processes. They suggest that the zone consists of a relatively gentle upslope section and a steep downslope end (Fig. 5.7) and that these zones are separated by 'transport zones'. By an unspecified mechanism this 'perturbation' of the bed profile will cause others downstream. They envision slugs of sediment, discharged suddenly into the channel, moving as discrete but diffusive waves downstream. Quantitative support for this latter concept comes from the thorough analysis of bedload movement in the East Fork River, Wyoming (Meade et al. 1981, Emmett et al. 1983). Measurement of bed elevation at forty cross-sections showed that coarse sand and fine gravel discharged from Muddy Creek, a tributary of the East Fork, moved over an immobile, coarse, gravel bed at velocities of about 10-20 m/day between distinct storage areas in the channel spaced about 500-600 m apart. Each storage area could hold nearly all of the annual yield. Radically different relationships between water discharge and bedload transport were found above and below storage areas. Weir (1983) has explored a mathematical explanation for bedload wave movement in the East Fork River, whereas Pickup et al. (1983) have modelled the downstream movement of sediment from copper mines in New Guinea as a moving wave. Much larger scale seasonal storage of sediment has been proposed to explain observations on the Amazon (Meade et al. 1979) and Orinoco (Meade et al. 1983a, b) rivers.

A single event, such as a major storm that causes massive landsliding in a basin, may lead to dramatic changes in channel behaviour and load which persists for long periods of time (Schumm and Parker 1973). Such changes may precipitate others, such as the aggradation and flooding-induced activation of earthflows described by Kelsey (1980) along the Van Duzen River in northern California. Ichim et al. (1980) discuss the effects of periodic flood flows on stream bed elevation in Romanian rivers. Pearce and Watson

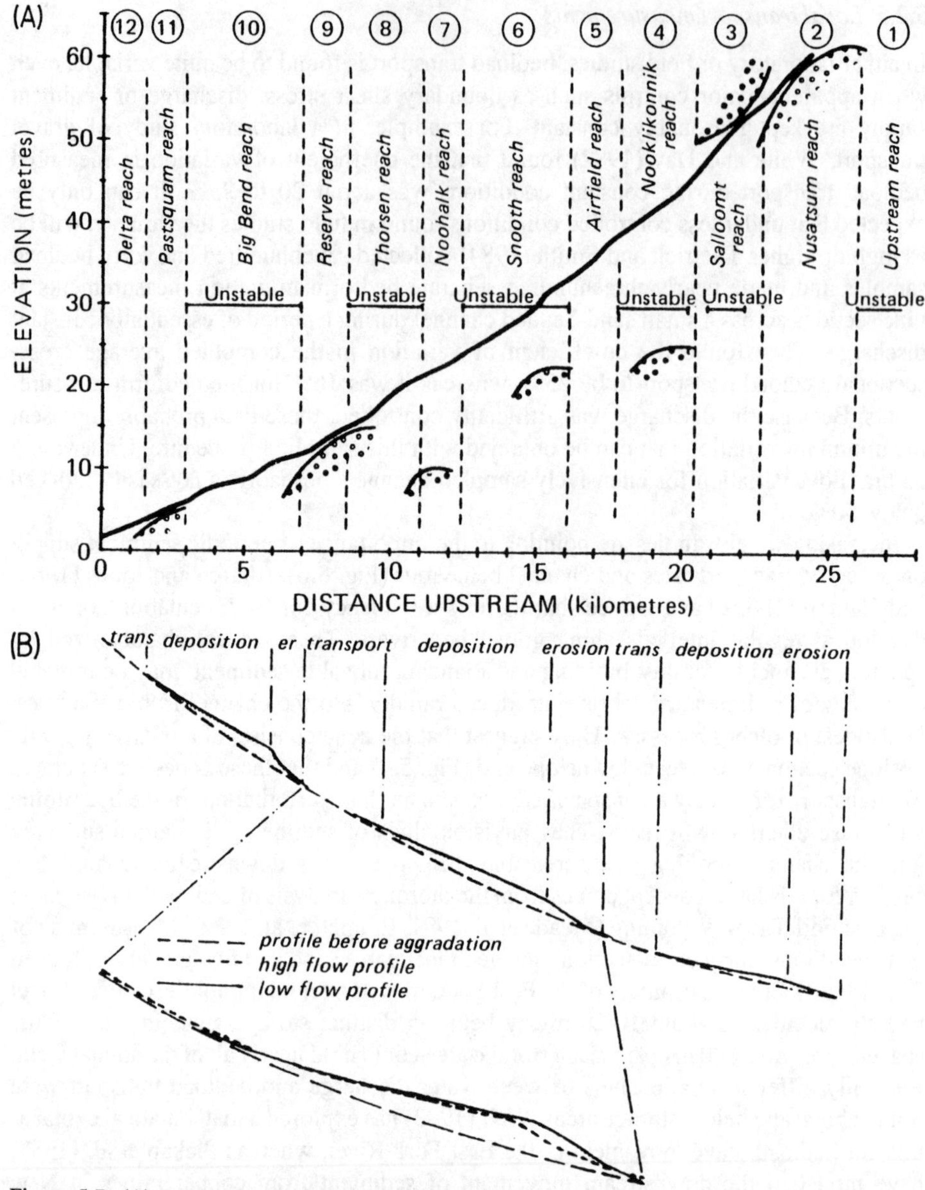

Figure 5.7. (A) Pattern of unstable 'sedimentation zones' associated with tributary alluvial fans (curved, textured symbols) along the Bella Coola River, Canada (from Church 1982). (B) Changes in the longitudinal profile of a river associated with aggradation in 'sedimentation zones' (from Church and Jones 1982).

(1983) examine the changes in channel storage after massive landsliding in a small basin and concluded that the effects will persist for decades, and Lisle (1981, 1982) has found that streams in northern California and southern Oregon aggraded for 5 to 15 years after a major flood and landsliding event in 1964. A major summary of the numerous studies undertaken in northern California, including detailed documentation of changes in sediment storage after the 1964 event, will appear as a US Geological Survey Professional Paper edited by Nolan, Marron and Kelsey. Within the context of this review, the main point to be made here is that changes in sediment storage may greatly affect the sediment transport rate observed at any particular section of a river. This point is pursued in some depth by Andrews (1979a, b).

Recent studies have pointed to the effect on local sediment transport rates of changes in sediment storage in debris jams (e.g. Mosley 1981, Keller and Tally 1979) and upstream riffles or pools (Ashida et al. 1981, Jackson and Beschta 1982). Cutoff of upstream meanders may cause either a local downstream increase or decrease in sediment load. With the reduction in trapping in the American West, beaver dams are once again becoming common place and are causing drastic changes in sediment load and stream flow in many small streams.

Finally, on the scale of a few channel widths, sediment transport, particularly bedload, may be highly variable both in space and time due to local processes. The interaction of the flow with channel curvature and bar-pool topography leads to asymmetric boundary shear stress distribution and steep cross-stream bedslopes which in turn may cause systematic cross-stream variation in bedload transport rates through a reach of channel as shown in Figure 5.8A (Dietrich et al. 1979, Dietrich et al. 1983, Dietrich and Smith 1984, Dietrich et al. 1984, Leopold 1982, Rakoczi 1983). Guanghua et al. (1983) have found that 50 to 70% of the bedload transport in the Yangtze River occurs in less than one-third the channel width. Sand-bedded streams, and many gravel ones are covered by mobile bedforms which cause bedload transport at a position along a cross-section to vary from zero to large values (Fig. 5.8A). Even in gravel-bedded streams where dunes or ripples were absent, we (with Leopold) have observed distinct concentrations or bedload sheets roughly one grain in diameter in thickness which moved past measurement points causing several orders of magnitude change in bedload transport rate. Southard et al. (1981) describe similar gravel bedload concentrations that form when a chute incises into a lobe in a braided channel and discharges a slug or moving low-relief patch of sediment. Church and Jones (1982) describe 'diffuse gravel sheets' that are gravel waves typically a few grain diameters thick and which appear to accumulate into bars at points of flow divergence. Bluck (1982) has found sedimentological evidence for such sheets. At low flows in gravel-bedded streams, sand tends to travel as an intermittent bedload and suspended load and can form into distinct streaks (Rachocki 1981) travelling across the coarser gravel (see also Jackson and Beschta 1982). Klingeman and Emmett (1982) discuss further other sources of temporal variation in local bedload transport rate, particularly those associated with the initiation of motion of the armour layer in gravel-bedded streams (see also Gomez 1983). Jaeggi and Smart (1982) commented on the paper by Klingeman and Emmett and state that local transport rates are controlled by upstream filling, and scouring of pools.

Ying (1983) found he could characterize the variation in gravel bedload transport for single discharge values in the Minjang and Yangtze Rivers with a general distribution function. He attributed the variation to intermittent gravel surges resulting from many factors, most of which we have discussed above. From his analysis he points out that diversion works in hydroelectric stations should be designed to handle the tail of the distribution where the instantaneous bedload transport can exceed the average transport for that discharge by a factor of 2.6 times.

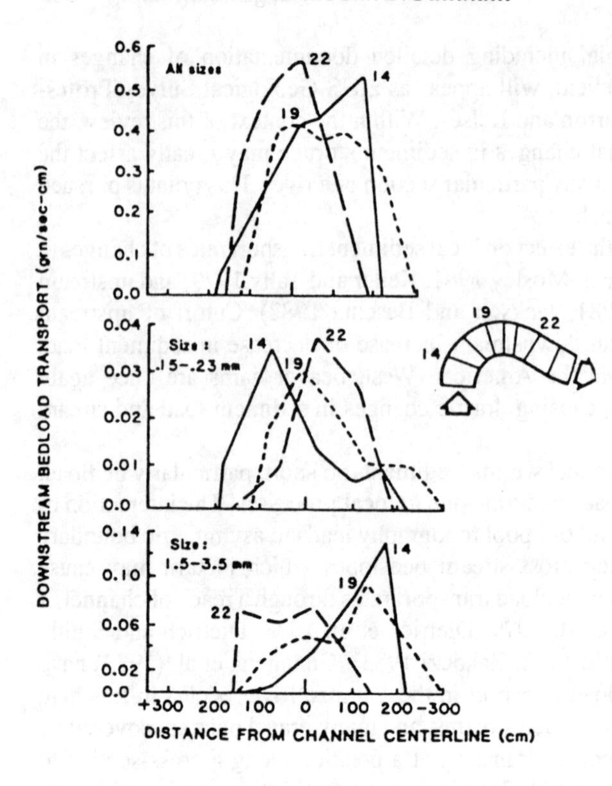

Figure 5.8. (A) Downstream variation in the bedload transport field for sediment of different sizes in the sand-bedded river, Muddy Creek, during a period of constant discharge and equilibrium bed topography (from Dietrich and Smith 1984).

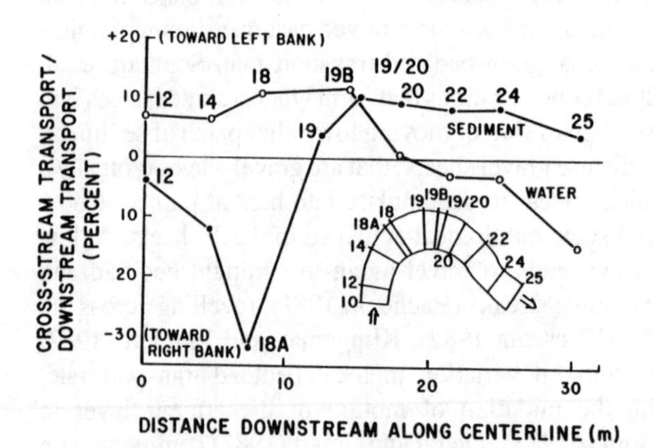

Figure 5.8. (B) Rates of cross-stream to downstream transport of bedload and water as a function of distance downstream through the bend depicted on the graph. Curve with solid circles represents the bedload. Geometric crossing to downstream bend is below section 22. Negative values are toward right bank which, in the upstream bend, is the inner bank. Measurements made during a period of constant discharge and equilibrium bed topography in a sand-bedded meander (from Dietrich and Smith, in press).

The many sources of variation in local sediment transport rate as illustrated in Figure 5.6 suggest that to obtain measurements of sediment transport rate (particularly bedload) useful for understanding channel morphology and behaviour, repeated sampling at a spot is required and much effort must be given to quantifying the controls on local sediment supply. Bedload measurements are becoming common place, but they are of little value unless the sampling density in time and space is sufficient to overcome inherent large variations found in natural streams.

We also wish to stress here that although it has long been recognized that sediment transport in a reach of river has both a downstream and cross stream component, these components have rarely been measured. Although probably small in most rivers, the cross-stream component may play a major role in adjusting channel topography in reaches with downstream varying boundary shear stress fields. For example, Dietrich and Smith (1984) measured the cross-stream and downstream components of bedload transport in a sand-bedded meander with equilibrium bed-topography and found a net cross-stream bedload that was toward the pool and was an average 10% of the downstream bedload transport (Fig. 5.8B). This relatively small cross-stream component allowed the zone of maximum bedload transport to shift outward through the bend and follow the outward shifting zone of high boundary shear stress, leading to the point bar equilibrium.

5.2.6 *Sediment load*

Measurement of suspended sediment discharge in rivers has become common place, and bedload measurements are rapidly becoming so. Still problems remain concerning sampling methods, including frequency of sampling, choice of instrumentation and calibration of instruments. These topics are extensively addressed in the 1981 Proceedings of the Florence Symposium on Erosion and Sediment Transport Measurement (IAHS-no. 133) in which 30 papers are presented on the issue of measurement of sediment transport in rivers. We defer to the useful reviews contained there.

In recent years hyperconcentrated flows (Newtonian fluids with very high sediment concentrations) have attracted considerable interest because of the dam construction on rivers in China with very high suspended loads (Zhengying 1983) and because of the tremendous release of sediment into rivers after the eruption of Mt. St. Helens in Washington State, USA (USGS Prof. Pap. 1250). In the recent River Sedimentation meeting in China, six papers concerning the mechanics of hyperconcentrated flows were reported by various collaborative Chinese groups involving 15 individuals in total. These papers examine the rheology, turbulent structure, roughness and settling velocity properties of hyperconcentrated flows. At the annual meeting of the American Geophysical Union in San Francisco, December, 1983 (EOS Vol. 64, No. 45) five papers on mudflows and hyperconcentrated flows associated with the Mt. St. Helens eruption were given in a special session entitled: Transport processes of excessive loads. These papers primarily focussed on distinguishing mudflows from hyperconcentrated flows and on application of proposed flow theories to field observations.

5.3 BED TOPOGRAPHY

5.3.1 *The scour hole and lobe front bedform*

In recent years a clearer understanding has emerged of the geometry of stream beds and the processes that dictate bed morphology. The riffle-pool sequence is now being seen in many instances as a descriptive division of the same basic bedform (Fig. 5.9). This bedform (referred to as a bar unit in Fig. 5.9) consists of an upstream narrow scour hole (pool) that widens and shoals to a lobate, bar crest. If the bed becomes immobile when the

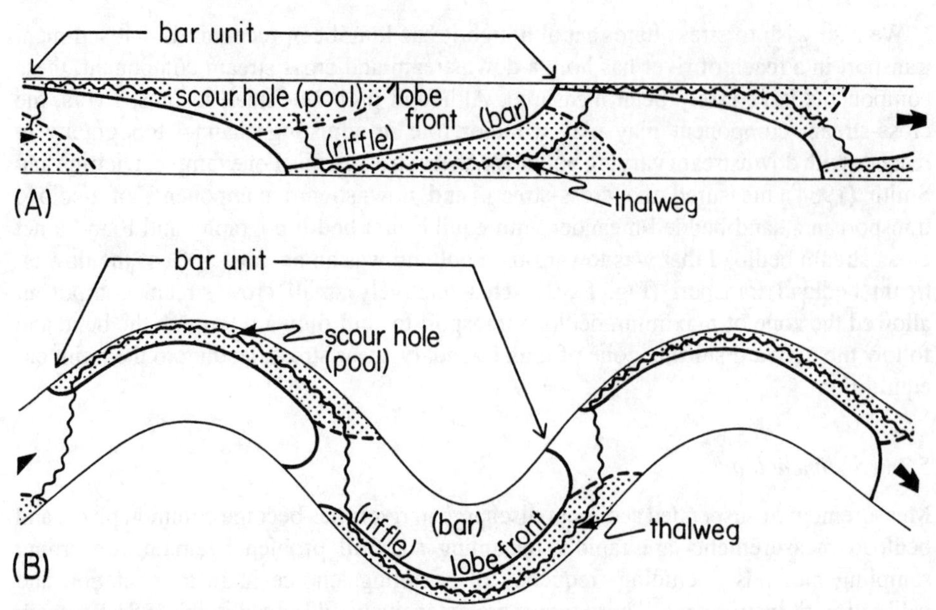

Figure 5.9. Fundamental large-scale bedform in rivers. (A) The bar unit consists of an upstream scour hole (shaded region) which widens and shoals to an obliquely oriented lobe front. The terms pool and riffle define specific parts of the same bedform (illustration modified from Parker and Peterson 1980). (B) In meanders, the bedform extends from the upstream pool to the downstream bar. Unlike in (A), the bedforms in this case do not migrate downstream if the banks remain stationary.

stage declines then at low flow water will spill from pool to pool across the oblique front of the lobe. This is the area normally referred to as the riffle. Numerous authors have proposed the distinct scour-lobe (pool-riffle/bar) morphology as the fundamental large-scale bedform in rivers (Kinoshita (1961) as reported in Chang et al. (1971), Yamaoka and Hasegawa (1984), Parker and Peterson (1980), Ashmore (1982), Church and Jones (1982), Jaeggi (1984), Ferguson and Werritty (1983)). From this perspective, the debate about the hydraulic properties of pools versus riffles seems of less significance, because it is the hydraulics of the whole bar unit, not just artificial subdivisions of the same unit, that should be examined.

Although few laboratory or field studies have investigated the detailed pattern of flow and sediment transport in a straight channel with alternate bars it is apparent that the bars force net cross-stream discharge of water and sediment and strong spatial variations in boundary shear stress (Kinoshita (1961) in Chang et al. (1971, Fig. 3), H. Ikeda (1983, Figs. 3.30 and 4.2), S. Ikeda (1984, Figs. 1 and 2); also see discussion p. 1561-1563 in Parker and Peterson (1980) and Leopold (1982)). As mentioned earlier (Fig. 5.3) the Smith and McLean model (1984) predicts strong skewing of the flow and development of secondary circulation in straight channels with alternate bars. The predicted pattern is much like that for curved channels with bar-pool topography and, as in river bends, the equilibrium alternate bar topography probably becomes established when outward transport over the top of the bar causes net transport away from a zone of decreasing boundary shear stress and into an increasing one in the deeper water (see Dietrich and

Smith, in press). Also, skewing of the flow and the development of secondary circulation in the pools contribute to bank erosion.

Alternate bars have been shown both in laboratory flumes (e.g. Wolman and Brush 1957, Ackers and Charlton 1970, Parker 1976, Plate 2) and in field study (Lewin 1976, Parker and Andres 1976) to be precursors to the development of meandering. Olesen (1984) has questioned the validity of these observations and of the previous mathematical models that draw this conclusion (e.g. Parker 1976, Engelund and Skovgaard 1973, Fredsoe 1978) because his theory suggests that when the banks are relatively unerodible the bars would migrate downstream too quickly for erosion at the banks to alter significantly the channel form. Such a condition may be met in the reach of Colorado River studied by Leopold (1982), but not in channelized, unprotected banks of the rivers studied by Lewin (1976) and Parker and Andres (1976). Field and laboratory studies listed above have shown that the alternate bars migrate slowly downstream, but as meandering develops the bars become 'locked' in place in the bends and only move as the channel migrates. The same scour-lobe or pool riffle/bar bedform is, in a sense, wrapped around the meander, and, surprisingly, the bedform extends from the upstream end of the pool in a bend to the downstream edge of the point bar in the next bend downstream (Fig. 5.9B).

The origin of the instability in a plane bed that leads to alternate bar topography is still not completely understood. Some would argue that due to boundary effects the flow is inherently unstable and longitudinal oscillations in the downstream velocity field will develop (e.g. Einstein and Shen 1964, Leopold and Wolman 1957, Shen and Kamura 1968, and references in Parker 1976). Alternatively it has been proposed that sediment movement, leading to slight bed undulations plays a necessary role in the initial bed instability (e.g. Parker 1976 and references therein). Recent experimental work by Nakagawa and Hotsuta (1984) shows that, as Nakagawa (1983) had previously suggested, differences in boundary shear stress between the bed and banks in wide, shallow channels influences the development of a weakly sinuous downstream flow in a trapezoidal channel. The wavelength of the oscillation was 2.5 to 6.3 channel widths, comparable to alternate bar spacing. There is still a need in Nakagawa's or others' experiments to show by detailed, direct measurement the connection of the flow instability with bed material movement and the growth of alternate bars. Also, some theories (i.e. Parker 1976) suggest the bar spacing scales with depth and boundary roughness as well as channel width (see S. Ikeda 1984). Ideally, these problems would be tackled by field experimentation.

It has been proposed that as a bend like the one in Figure 5.9B grows in amplitude, the center line path between crossings (inflection points) will become sufficiently long that the alternate bar instability will lead to the formation of another scour hole and lobe within the bend (Keller and Melhorn 1978, Hooke and Harvey 1983). Hooke and Harvey report observations on the River Dane which suggest that the path length may vary with channel curvature, tending to be longer for more strongly curved channels. The second pool that forms in these bends is also along the convex bank rather than near the concave one, unlike alternate bar topography produced in straight channels. Mathematical modelling by De Vriend and Struiksma (1984) now suggests that in many cases the secondary pool in a bend arises from an oscillation in the flow and sediment transport caused by the rapid curvature change at the entrance to the bend. Flume experiments at the Delft Hydraulics Laboratory in a bend with constant curvature and long straight entrance and

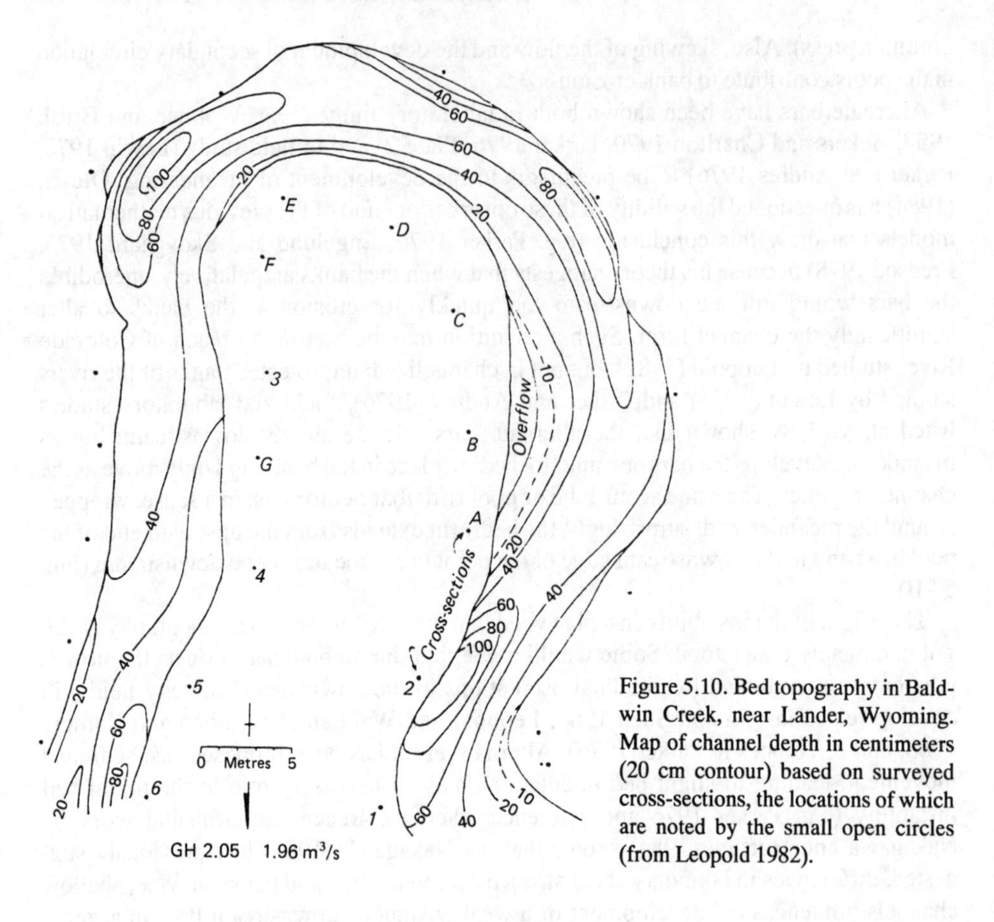

Figure 5.10. Bed topography in Baldwin Creek, near Lander, Wyoming. Map of channel depth in centimeters (20 cm contour) based on surveyed cross-sections, the locations of which are noted by the small open circles (from Leopold 1982).

exit reaches produced a bed with two pools spaced about 12 channel widths apart. The general bed topography was predicted very well with the De Vriend and Struiksma model. Other examples of constant curvature bends with straight adjoining sections having multiple pools are readily found (e.g. Kondrat'ev 1968, Kennedy et al. 1984). The De Vriend and Struiksma model predicts that in very long bends of constant curvature the 'transitional effects' due to channel curvature change will be damped out. The observations reported by Engelund (1975) regarding his experiments in a circular annulus with a mobile bed would suggest otherwise. He found about five to six 'scour holes' over the length of the flume, equivalent to an average spacing of five channel widths. So it appears that De Vriend and Struiksma have identified an important curvature change-induced flow oscillation that would produce a secondary pool, but that other flow instabilities may also play a role in formation of multiple pools in a bend.

The application of these insights to natural river bends is difficult. For example, Figure 5.10 shows the bed topography in Baldwin Creek, a gravel-bed river near Lander, Wyoming (Leopold 1982). Four pools are readily identified. The pool downstream of the bend apex may be caused by the mechanism proposed by De Vriend and Struiksma, but there is also a local curvature maximum at each site. Also the channel at first widens and then rapidly narrows at the more upstream pool in the central bend. With erodible banks, the tendency for a second pool to form may cause local acceleration of erosion and the formation of a local radius of curvature minimum which would enhance the pool formation. Bank material variations further complicate the problem. Nonetheless, careful field experiments will help considerably in examining the application of these concepts to explaining the development of alternate-bar bed topography.

Of course, other large scale bedforms are found in rivers and a plethora of terms has been proposed. The most interesting recent discussion of this matter is in the paper by Church and Jones (1982) and in the discussion papers that followed it, particularly that provided by Ashmore. Church and Jones point out the difficult problem of comprehending the effects of heterogeneous sediment size and fluctuating discharge on bedform morphology. Japanese scientists generally follow the Kinoshita classification system of single row alternate bars, double row alternate bars and bars with higher mode (S. Ikeda 1984). Single row alternate bars can be distinguished by how successive scour-lobe forms connect (see Fig. 4, Yamaoka and Hasegawa 1984). Thus the scour-lobe bar unit is seen as a genetic template to build most bed features. In contrast, Gergov (1983) briefly describes a classification system used in the USSR and Eastern European countries.

5.3.2 *Medial bar and smaller bedforms*

The other general bedform that can be included in all classifications is a medial bar. It may be the downstream lobe from a scour hole (Ashmore and Parker 1983) or it may be a localized bump associated with deposition (Leopold and Wolman 1957). Such bars are associated with braided channels. These bed features have received much less detailed study (Bluck 1976, Southard et al. 1981) but as more work focusses on understanding braided channels the mechanics of formation and sediment transport over these features will be more fully revealed.

Smaller scale bedforms, such as ripples, dunes, and antidunes, are usually found in sand-bedded streams and are rarely observed in coarse gravel ones (but see Gustavson 1978, Whittaker and Jaeggi 1982). There is an extensive literature, both empirical and

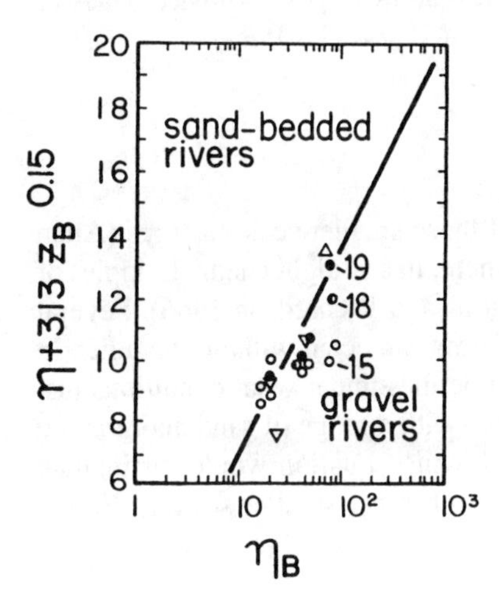

Figure 5.11. Comparison of the Jaeggi equation to distinguish the condition for alternate bar formation (solid line) with flume and field data. Open and closed circles are data from the flume experiments by Ikeda (1983) for conditions of well developed alternate bars and antidune or plane bed, respectively. Split circles are transitional data points. Numbers refer to 'run nr' in Figure 5.12. Triangle is the predisturbance condition on the river studied by Lewin (1976). The inverted triangles represent the range of data reported by Andrews (1984) for gravel-bedded rivers in Colorado.

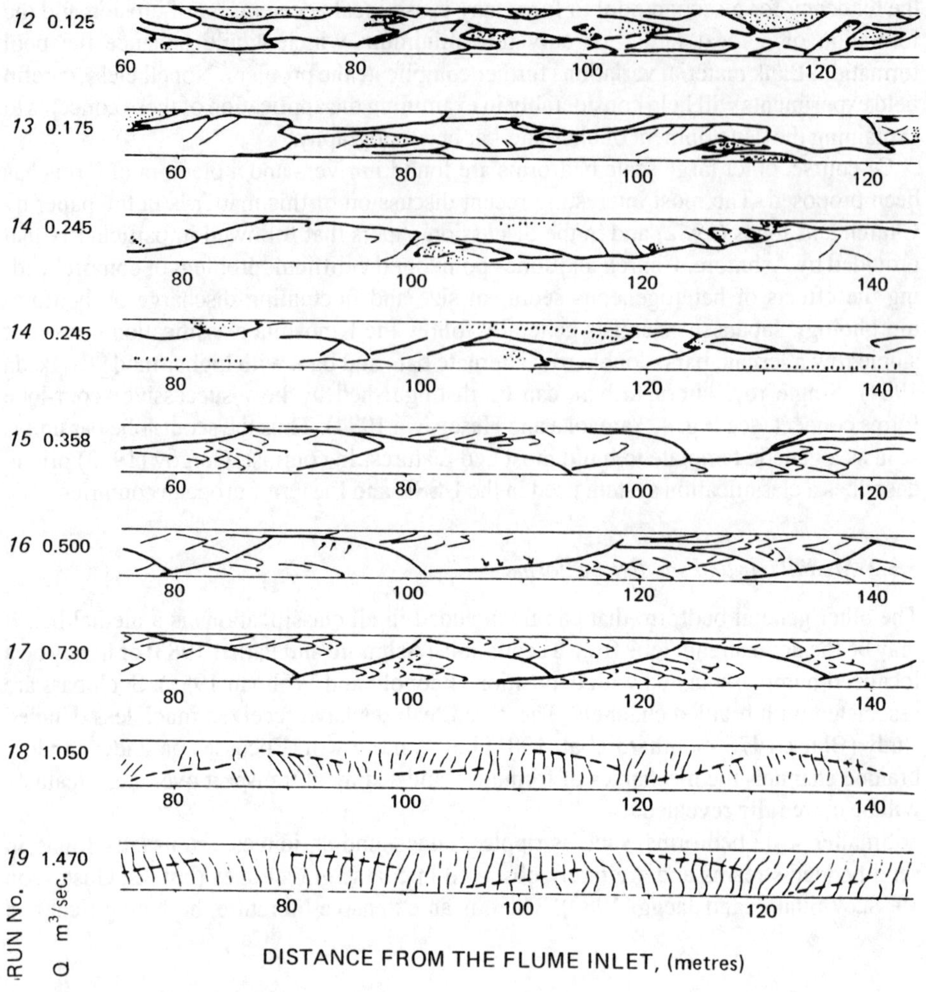

Figure 5.12. Changes in shape of bed configurations with increasing discharge in a 4 m wide flume with a bed slope of .01 and a median particle size of about 6 mm (from H. Ikeda 1983).

theoretical, on factors controlling size and shape of these smaller-scale features (Allen 1982). Some flume studies have found conditions where, in a straight channel, ripples or dunes develop on alternate bars (Ackers 1982, Simons and Richardson 1966). Several flume studies indicate that large or small-scale bedforms can occur without the other. In an attempt to establish an empirical expression that defines under what conditions just alternate bars formed Jaeggi (1984) did flume studies with a range of sand and granule size sediment. He found that in most cases the following equation would distinguish between channels with alternate bars and channels with other smaller scale bedforms (dunes, antidunes, plane bed, diagonal bars):

$$\frac{hS}{H_c\left(\frac{ps}{p}-1\right)} = 2.931n = \frac{wS}{H_c\left(\frac{ps}{p}-1\right)D_m} - 3.13\left(\frac{w^{0.15}}{D_m}\right) \quad (5.13)$$

where D_m equals the mean grain size, H_c equals the dimensionless critical Shields number $\left(\frac{\tau_c}{(\rho_x - \rho)gD_m}\right)$, w is the channel width, and other terms are defined as before. Equation 5.13 is plotted in Figure 5.11 along with flume and field data not analyzed by Jaeggi. The data from Ikeda's (1983) experiments agree fairly well with Equation 5.13. The symbols labelled with numbers refer to the run number sequence shown in Figure 5.12. Ikeda found that as he systematically increased discharge but held width, slope and particle size constant, the bedform pattern changed systematically from braided to alternate bar to antidune. Inspection of data that plot to the right of the line in the alternate bar field reveals that they also plot in the lower and upper plane bed stability field in bedform phase diagrams (p. 340, Allen 1982). Jaeggi reports observations of plane bed, however, that lack alternate bar formation. It is difficult to find field data to plot on Figure 5.11 and only a few examples are shown. Sandy braided and meandering rivers ranging in size from Muddy Creek (Fig. 5.8A), to the Platte and Mississippi Rivers plot well above the line, despite having well developed alternate bars. Coarse gravel rivers will tend to plot below the line. The river studied by Lewin (1976) plots just slightly above the line when it was channelized and probably moved below the line as the alternate bars developed and the channel widened. More work, both theoretical and field measurements, is needed to establish the usefulness of the distinction proposed by Jaeggi.

5.3.3 *Obstruction-controlled bed topography*

As a final comment on large-scale bedforms, most natural rivers encounter obstructions either along their banks or beds that generate local scour-lobe features, lock in place migrating bars, and cause along channel variations in bank erodibility, hence the planform curvature effecting size and location of bars. Location, size and stability of pools and riffle barriers in streams has also become an important issue of stream ecology particularly with regard to fish communities (e.g. Bjorn et al. 1977, Power and Mathews 1983). Although obstructions such as bedrock outcrops, trees or boulders can have a substantial local effect on channel morphology, at least one study suggests that on average the alternate bar spacing remains similar to that found in laboratory studies (Keller and Melhorn 1978). Keller and Swanson (1979) suggest that the influence of 'large organic debris' (isolated timber debris, log jams, root wads) changes with increasing channel size downstream. The sequence consists of a random distribution of debris in small steep headward channels, distinct accumulation zones which may effect the entire channel width in intermediate streams, and scattered large debris having little influence on channel conditions in large rivers. As part of her Ph. D. dissertation research at University of California, Santa Barbara, Ann McDonald has performed debris jam removal experiments in small streams in northern California to document their role in channel morphology, flow velocity and sediment storage. Lisle and Kelsey (1982, 1983) are currently documenting the role of bedrock outcrops and boulders on generating pools and bars and on preventing bar migration in streams in northern California. They are pursuing the problem of predicting the size, and shape of the scour-lobe topography for a given size and deflection angle of bedrock obstruction. Gallinatti (unpubl. manuscript) has examined this problem in a small channel inundated with sediment after the massive eruption of Mt. St. Helens in 1980.

5.3.4 *Hydraulic geometry*

One of the most fundamental issues in channel topography is what controls the mean width and depth of a reach of channel. In the last few years two camps, both building upon the seminal work of Leopold and Maddock (1953), have emerged in this area. Following the proposition of Langbein (1966) and Langbein and Leopold (1964) that the hydraulic geometry relationships can be predicted from minimum variance and general energy expenditure concepts, several researchers have derived expressions for channel geometry (e.g. Sang and Yang 1980, Chang 1979) and have used these concepts for several other problems.

The alternative approach, based on the work of Lane (1953) derives the hydraulic geometry relationships using specific assumptions regarding distribution and magnitude of boundary shear stress at channel forming stages. Parker (1978a, b) suggests that the cross-stream variation in boundary shear stress in sand and silt channels requires that a balance must be struck between the tendency for sediment to be carried as bedload toward the centerline of the channel where the shear stress is highest and the tendency for suspended sediment to settle out on the inclined bank where the shear stress is lowest. In contrast, in gravel-bedded streams he postulates that the cross-channel shape is one that leads to shear stress declining monotonically from the bed and up the bank such that at bankfull discharge the transition location from bed to bank experiences shear stress equal to the critical value for the sediment size present. Osterkamp et al. (1983) make essentially the same assumption as Parker, but take a more empirical approach to deriving hydraulic geometry relationships. The field studies by Andrews (1982, 1984) have provided important support for the Parker models, by describing bank erosion and bed deposition processes in a gravel bed river with a sand cover and by reporting data for gravel bed rivers that indicate initial motion occurs at slightly less than bankfull discharge. More field work is still needed which investigates the pattern of bed and bank boundary shear stress and sediment transport in gravel bedded reaches. The main difficulty will be to make a bank shear stress measurement and to evaluate initial motion values for sediment on an inclined slope (S. Ikeda 1982) and for cohesive sediments (see Thorne 1982).

5.4 PLANFORM CHANNEL GEOMETRY

In the past few years the study of river planform geometry and the factors that control it has become one of the most productive and exciting areas of fluvial geomorphology. A rich interplay between theoretical analysis, field and laboratory measurements and numerical simulations has begun and promises to lead to strongly focussed questions and answers more firmly based on an understanding of the complex interactions of flow, topography and sediment transport that controls channel form. Here we summarize some recent developments regarding meander planform and then discuss briefly research on braided channels. For a discussion of channel planform classifications see the reviews in the papers by Brice (1984), Schumm (1981) and Church and Jones (1982).

5.4.1 *Meandering rivers*

Langbein and Leopold (1966) made a compelling argument that highly symmetric meanders (like that drawn in Fig. 5.9) are well represented by a sine-generated curve. Although they acknowledged that most rivers were not perfectly symmetrical, they argued that the sine-generated curve represents a path that minimizes the variation of important hydraulic properties and as such is the ideal or 'most perfect' form a river would take given no variation in bank erosion properties. Because of the high degree of planform variation in natural rivers some researchers have maintained that no one simple, general form exists (Brice 1973, 1984). On the other hand, Ferguson (1973, 1976) showed that the sine-generated curve was similar to that proposed by Fargue and that random perturbation in a sine-generated curve path would generate the more irregular pattern found in nature.

At the same time, research by individuals from many disciplines began to focus on developing an understanding of the processes controlling the flow field and bed topography in a single bend of a meander (see review in Allen 1978, Dietrich and Smith 1983, 1984). A particularly important mathematical model was proposed by Engelund (1974) that yielded a prediction of the bed topography developed in the flume studied by Hooke (1975). A significant feature of Engelund's 'second approximation' is that it provided the basis for predicting the downstream change in the velocity field as a consequence of bank curvature and bed topographic effects. Although the Engelund model is mathematically and physically incorrect (see Kalkwijk and De Vriend 1980, Dietrich and Smith 1983, 1984, Smith and McLean 1984), it can be used to simulate reasonably well flow and topographic features of river bends (Bridge 1984).

In a series of two important papers, Ikeda et al. (1981) and Parker et al. (1982) found that if they used a flow model quite similar to the Engelund one coupled with a simple assumption about the relation between near-bank velocity and erosion that they could predict approximately the wavelength of river meanders and a characteristic 'skewing' or asymmetry to the planform (Fig. 5.13). The asymmetry arises from the cross-stream shifting of the high velocity core from the convex bank towards the concave one downstream of the bend apex. This outward shifting is well established in field studies (e.g. Leopold and Wolman 1960, Dietrich et al. 1978, Dietrich and Smith 1983, De Vriend and Geldof 1983, Bridge 1984). More complete flow models (Smith and McLean 1984, De Vriend and Geldof 1983) and flume experiments (C. Yen 1970) indicate that the outward shifting is primarily due to the bar-pool topography associated with curved channels (Fig. 5.4).

Following up on the Parker et al. (1982) findings, Parker and his graduate students (Parker et al. 1983) have reopened the question of whether meandering rivers, particularly larger amplitude ones, develop a characteristic form, or 'ideal state', which can be maintained as the river migrates. They performed a stability analysis of the equation developed by Ikeda et al. (1981) and found the following equation for a constant form meander:

$$H = H_0 \sin 2\pi\frac{S}{M} + H_0^3\left[\frac{1}{192}\sin\frac{6\pi S}{M} + \frac{\sqrt{s(A+F^2)}}{128}\cos\frac{6\pi S}{M}\right] \tag{5.14}$$

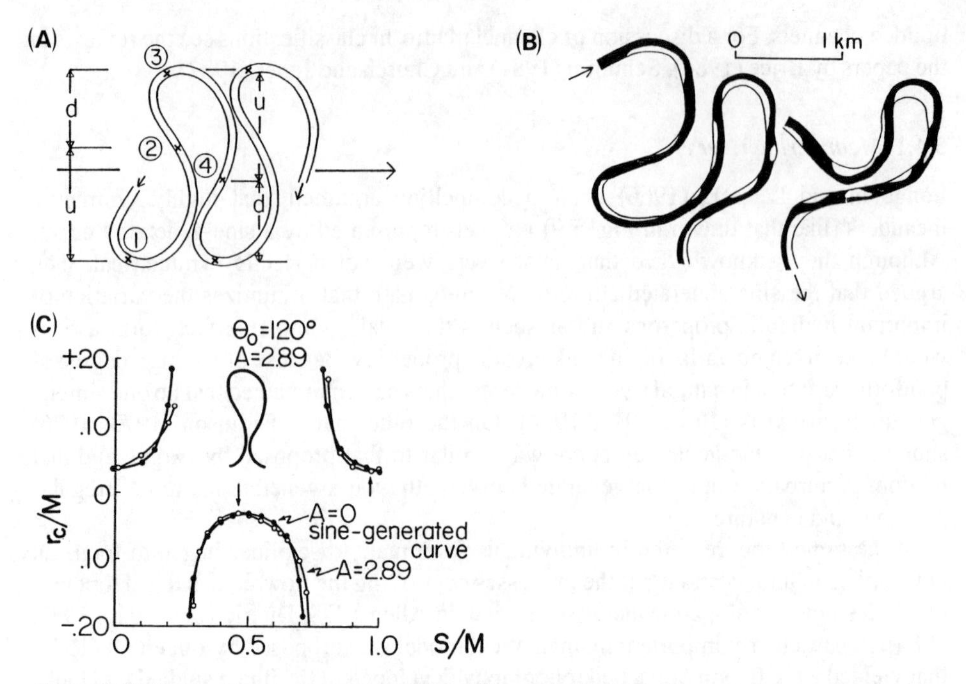

Figure 5.13. Shape of large amplitude bends. (A) Form predicted by Beck (1984) using the Ikeda et al. theory. The crosses mark the successive locations of the position of minimum and infinite radii of curvature through the meanders. The letters u and d refer to upstream and downstream limbs of a meander starting at a radius of curvature minimum (see Equation 5.16). The circled numbers are referred to in the text. (B) Tracing of two geometrically similar reaches of the Beatton River, Canada, shown in a photograph in the paper by Nanson and Hickin (1983). Note the multiple bars (unshaded portions) within these bends. (C) Comparison of the shape of large amplitude bends with the scour factor, A, in Equation 5.14 set equal to 2.89 and with A = 0.0 (sine-generated curve). Vertical axis is the ratio of local radius of curvature to total wavelength of the meander. The horizontal axis is the ratio of distance through the meander at the upstream radius of curvature minimum to the total wavelength of the meander.

Here H is the angular difference between the centerline orientation and the downvalley axis, and H_0 is the maximum value, located at the crossing; S is the distance downstream along the centerline path starting in this case at the upstream bend apex (Fig. 5.13) and M is the along channel wavelength (1 to 3, Fig. 5.13); F is the Froude number $\left[\frac{u_0}{\sqrt{gh_0}}\right]$ for a straight channel with the same width, roughness and discharge; A has been called the 'scour factor' and can range from less than 3.0 to as high as 30 (Parker et al. 1983).

The scour factor is the crucial parameter in Equation 5.14. It arises from the incorrect (Dietrich and Smith 1983, 1984), but nonetheless useful, assumption that at equilibrium the cross-stream component of the weight of a particle on a sloping point bar surface is exactly balanced by the inward component of drag caused by secondary circulation. With some approximations (see for example, Richards 1982, p. 208) this leads to

$$\tan\delta = A\frac{h}{R} \qquad (5.15)$$

where tan δ is the local slope of the point bar at a distance R along the radius of a local arc fitted to the channel planform, and h is the average channel depth. Although very approximate, Equation 5.15 and others similar to it have proven useful in estimating point bar slopes (e.g. Bridge 1984). In such estimations, however, A is freely adjusted to give a best fit. In Equation 5.14 it is crucial because, in most cases, F^2 is small, and it is the assigned value of A that determines the degree of asymmetry and the sharpness of the curvature. Large values of A yield a more asymmetric bend with lower radius of curvature minimum (see Fig. 15, Parker et al. 1983) Parker et al. (1983) report that $A = 2.89$ is typical of rivers in Japan and in Figure 5.15 we compare Equation 5.14 with $A = 2.89$ and $A = 0.0$, the latter giving the sine-generated curve. The effect of positive values of A in this case is to shift the position of minimum radius of curvature about .05S/M, roughly one channel width upstream from the halfway position and to reduce the radius of curvature at the apex and much of the remaining bend (leading to what Parker et al. (1983) have called a 'fattened' appearance to the bend). The difference between the two idealized curves is rather subtle and indicates the difficulty of attempting to confirm Equation 5.14 with field data. Parker et al. (1983) point out in fact that only fairly large amplitude meanders differ noticeably from the sine-generated curve (the first part of Equation 5.14).

Beck (1984) has used the Ikeda et al. (1981) equations to simulate meander evolution and to predict channel migration along a natural river (Beck et al. 1984). His numerical experiments indicate that downvalley migration systematically diminished with increasing sinuosity (or bend amplitude). As expected by the dominance of the sine-generated curve term in equations at low sinuosities, Beck found the radius of curvature to width ratio to reach a minimum of slightly less than 2.0. Langbein and Leopold (1966) have shown that at lower sinuosities the sine-generated curve should give consistent values of wavelength to radius of curvature values of about 4.6, which if the wavelength is ten times the channel width gives a radius of curvature to width ratio of 2.2. Beck's simulations suggest there is a critical wavelength below which wave amplitude diminishes and above which the bend grows and eventually cuts off as suggested by Parker (1984). In all cases the meanders migrated downstream.

The work by Parker and his colleagues has generated considerable interest. Some work by others has already been published, more is currently being pursued. Carson and La Pointe (1983), working at first independently of Parker et al.'s findings, analysed meander planform using the simple statistic

$$z = 100\, u/(u+d) \qquad (5.16)$$

where u and d are defined in Figure 5.13. For each loop, two traverses are used. If z is greater than 50 then the form is asymmetric with a convex downstream facing bank. Their analysis of 15 rivers yielded a median z of about 58, but with a larger fraction in most cases with $z > 65$. Brice (1984) has pointed out, the z measurement does not test directly for the symmetry of individual bends or loops, instead it indicates whether the full meander planform is asymmetric to the axis of the meander belt. Howard (1984) has proposed that bend asymmetry can be measured by taking the dis tance along the channel from (2) in Figure 5.13 to (3), subtracting from it the distance from (3) to (4) and dividing by the distance from (2) to (4). Kondrat'ev (1968) has also proposed a simple measure of bend asymmetry.

Howard (1984) has merged his own model for river meandering with that of Ikeda et al. (1981) to analyze by numerical modelling, the meandering over long reaches of channel. His numerical experiments indicate that low sinuosity channels are sinusoidal and these low amplitude bends translate rapidly downstream, but as they grow the translation reduces. He also explored incorporating the Nanson and Hickin (1983) field observations that migration rate appears to be a function of radius of curvature to width ratios. Ferguson (1984) has used the Nanson and Hickin curve to generate a purely kinematic numerical simulation of meander evolution. Skewing and double looping (Brice 1984) were induced by causing a lag between the location of radius curvature and the corresponding bank erosion in the calculation procedure.

The asymmetric issue is now being explored in flume experiments. Yamaoka and Hasegawa (1984) measured flow, bed topography in sine-generated and asymmetrical bends. Although an important start, interpretation of their results is difficult because it appears that the asymmetric channel had two radii of curvature minimums, one at about 0.45 S/M and the other at 0.7 S/M. Davies and Tinker (1984) performed experiments with a viscous fluid on an inclined glass plane and found that convex downstream meanders would develop at high amplitudes. From this they make what seems to be a great leap of faith: because such asymmetry patterns are found in special fluids on glass plates, asymmetry is the fundamental form of rivers. If one were to take a long metal chain, say with 5 mm diameter beads, and wiggle it back and forth at one end and hold it at the other, distinct convex downstream meanders will form. It is fair to say that the reason for such asymmetry has very little to do with river meanders.

The problem of river meander form is far from solved. Although Parker and his colleagues have produced consistent, predictive and practical results, there are troubling issues that must be dealt with. The fundamental equation upon which all their meander form modelling is based (Ikeda et al. 1981, Eq. 3b) can be shown to be physically incorrect because it includes only one of the two equally important convective acceleration terms (Dietrich and Smith 1983, Smith and McLean 1984). It should be possible to reformulate the basic equation and examine whether it matters to their results. Their basic equation predicts bank velocity and then they relate bank erosion rate to near bank velocity. In effect they have assumed a 'slip velocity' at the bank, which, although it may provide a good approximation, certainly is not physically correct. Confirmation of their predicted near bank velocity by comparison with field observation for bends of various shapes would add considerable weight to their findings.

Finally, a more subtle issue needs to be considered. Empirically it seems well established that low amplitude bends tend to be the most symmetrical. As bends grow the length along the channel centerline between original inflection or cross over points can become quite large. For example, if we represent the channel path with a sine-generated curve, then Langbein and Leopold (1966) have shown there are simple relationships between sinuosity (k), along valley wavelength (λ), and H_0:

$$M = k\lambda$$

$$k = \left[1 - \left(\frac{H_0}{2.2}\right)^2\right]^{-1}$$

If a straight channel with alternate bars spaced at 5 channel widths starts to meander, M grows proportionately to the increasing sinuosity, and H_0 tends to increase, so that by H_0 = 90°, M has reached 2λ. Hence the distance from crossing to crossing, which was originally 5 channel widths is now 10 widths. If the inherent instability of flow over a point bar in a curving channel reported by De Vriend and Struiksma (1984) is of general significance then within a bend of large amplitude, there should be a tendency for a secondary pool and bar to form. Such secondary features may contribute to local bank erosion and planform asymmetry. If the alternate bar instability results from a different mechanism, not related to curvature change, as suggested by Engelund's (1975) experiments, then in large amplitude bends there should be a tendency for multiple bars and pools. Figure 5.13B shows the bar distribution in two reaches of the Beatton River, each with roughly three complete asymmetric bends. Within these reaches the average bar spacing in channel widths is 8 (left channel) and 5 (right channel). Note particularly the near-alternate bar sequence in the downstream convex portion of the left channel.

In sum, planform asymmetry may arise from more than one cause: skewing of the flow, curvature induced bar-pool instability and alternate bar instability. The first cause has been explored, although with an incorrect model, and was shown to be important at large amplitudes. The other sources of instability may be unimportant, but this has yet to be shown. This problem still needs a vast infusion of field observations on flow, bank erosion, and planform evolution in river meanders that are designed to test explicitly these hypotheses.

5.4.2 *Braided channels*

Braided channel studies are still in a necessary descriptive stage, where beyond developing regime diagrams for conditions controlling channel pattern (Fig. 5.14), little in the way of quantitative predictions about form and behaviour can be made. Flow and sediment transport processes in braided channels are very complex, but field and theoretical experience gained in meandering river studies is now being applied to these channels. Here we briefly discuss some recent progress in three kinds of braided channels. Field experiments in braided channels which examine mechanics are greatly needed.

Field studies by Church, Neill and others (see Ferguson and Werritty 1983) in low sinuosity, wide, shallow, gravel bed streams that are weakly braided have yielded insight into the relationship between the scour-lobe topography described earlier (Fig. 5.9) and channel form. They propose calling these channels 'wandering gravel rivers'. Ferguson and Werritty show that the bed of one such river in Scotland is composed of migrating alternate bars with diagonal fronts of the form drawn in Figure 5.9A. These bars aggrade, migrate downstream, cause local bank erosion, and eventually experience avulsion on the side of the channel opposite the diagonal front (Fig. 5.15). The process repeats itself, leading to continuous channel instability. The study by Ferguson and Werritty was done over a five year period on a very active river and it is a particularly good description of the 'wandering gravel rivers'. It may be that such channel behaviour occurs where the scour-lobe topography cannot achieve equilibrium with the flow because the bar in the lobe front tends to grow out of the water rather than reach some equilibrium depth and diversion of flow toward the riffle leads to avulsion. Such conditions are more likely met in coarse, shallow gravel rivers. This conceptual model and the useful flume experiments by Ashmore (1982) and Ashmore and Parker (1983) will help considerably in detailed field experiments. Other sites where such processes are apparently taking place have already begun to be reported (Saucier 1984).

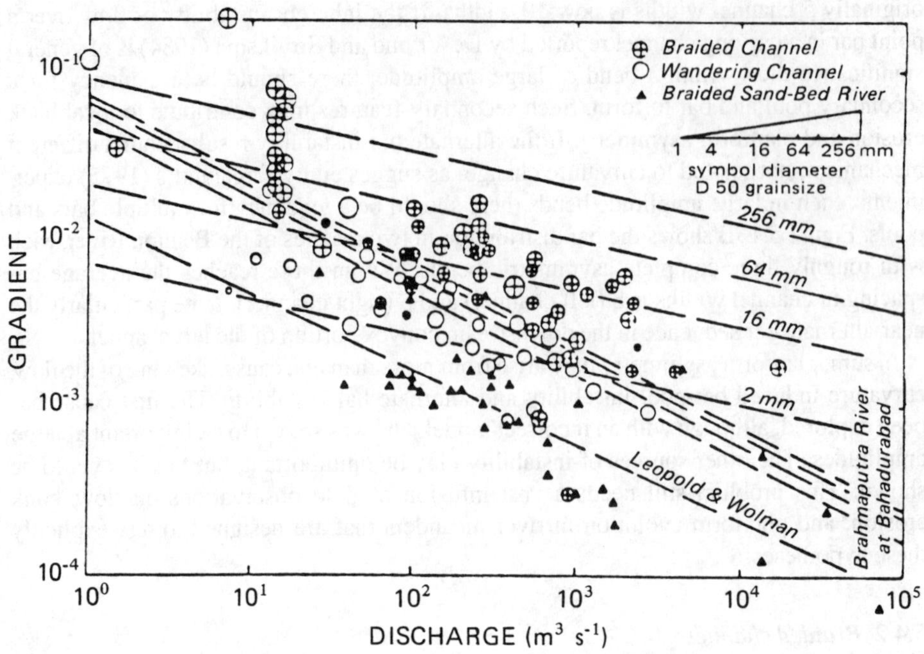

Figure 5.14. Discharge-gradient-grainsize plot of braided and nearbraided rivers (from Ferguson 1984).

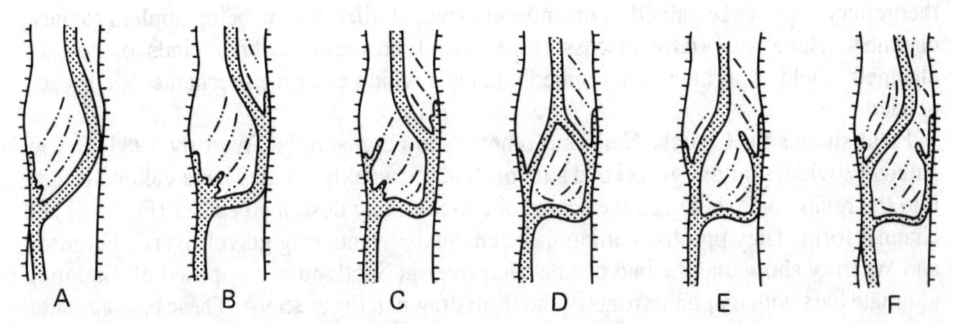

Figure 5.15. Sequence (A-F) of bar instability in a channel of low sinuosity (from Saucier 1984). This pattern appears to be typical of 'wandering gravel bed' rivers.

Sand-bedded braided channels may be quite different. The existence of large scale migrating bed forms (sand waves or linguoid bars) in sandy rivers leads to a braided pattern upon falling stage. Blodgett and Stanley (1980) describe the system of migrating linguoid bars found in the sand-bedded Platte River in Nebraska. During high discharge, the bed consists of an organized array of large scale migrating bedforms in a low sinuosity channel. They distinguish between staggered and nested arrays; however both types show a systematic organization of bars within each array. This contrasts with the fairly random distribution of bar forms found in the gravel bed 'scour and lobe' type of braided channels. During falling stage, the bars are partially dissected, with small delta lobes forming at the base of the dissecting chutes. The low flow channel pattern is a complex braided pattern which shows only minor resemblance to the ordered nature of the bed during high discharge.

Crowley (1983) also describes the bedforms in the Platte River, but concentrates on the downstream variation in bar type (defined by its high discharge form). The downstream variation is from a sinuous channel with braided alternating bars, through a reach with more coherent alternating bars, and leading eventually to the very coherent linguoid bars described by Blodgett and Stanley (1980). The nature of these transitions is not clear. Trenching of the bar deposits show foreset stratification due to avalanching lee faces in both the linguoid and alternating bars. These are capped by topset (planar) and festoon bedding created by vertical accretion from suspended load and bedload (dunes) respectively.

Smith and Smith (1984) describe the widening and braiding of the William River in Saskatchewan, Canada caused by the addition of large amounts of sand from an adjacent dune field. The bars in the river are similar to the linguoid bars of the Platte River and once again are large scale migrating bedforms, acting to transport a large sediment load at high discharge. Modification during low discharge leads to a complex braided bed configuration.

The geomorphic and sedimentary environment of anastomosing river systems has been addressed by Smith and Smith (1980) and Rust (1981). This pattern consists of a network of low gradient, narrow, deep, straight to sinuous, stable channels. The channels are separated by levees and vegetated wetlands which are built up by vertical accretion of silt and clay from overbank flooding. The form, sedimentology, process of formation, and geomorphic setting are all distinctly different from braided rivers. Smith and Smith (1980) have described three examples of anastomosed river systems in Canada. All three result from a rise in the local base level by alluvial fan deposition downstream of the anastomosed section. The rapid deposition of the fans occurred immediately after deglaciation, from 10 000 to 8000 years BP. This has led to deposition of fines in a delta-like environment associated with the decreased gradient. Deposition of fines and resulting vegetation growth has led to stable channel banks and deep narrow stable channels. Augering of the levees and intervening wetlands has demonstrated that the channels do not migrate significantly, resulting in vertical rather than lateral accretion of the floodplain. The anastomosed systems are separated from upstream braided systems by a transitional zone consisting of stable silt bars within a braided channel. Augering of these bars suggests that they are decreasing in size while aggrading upward. This is interpreted to be the result of downstream encroachment of the braided system. If the base level has stopped rising, the sediment supply from upstream will allow the braided system to prograde on to the anastomosed section by deposition of material leading to an increased gradient. This investigation demonstrates how an integrated study of channel morphology, sedimentology, and stratigraphy can provide significant insight into a specific fluvial system.

Rust (1981) has described an anastomosing river in central Australia. The existence of a similar channel morphology in an arid region indicates the generality of the form. Copper Creek has an anastomosing river system superimposed on a relict braided system. The cause of the change is not as clear as in the Canadian rivers. Although it may have been caused by a tectonic rise in the base level, Rust appeals to an increase in aridity as the driving mechanism. In any case, there was a change from sand and gravel deposition to clay and silt deposition, resulting in more stable banks and channels.

Field experiments that map boundary shear stress, flow and sediment transport fields in braided channels, in the context of an appropriate conceptual model about their overall behaviour are greatly needed. Research groups associated with K. Prestegaard, L. Leopold and J. Bridge are headed in that direction.

Regime diagrams like that shown in Figure 5.14, which express conditions requisite for braided versus meandering channels are of considerable theoretical and practical importance. Figure 5.14 represents the compilation of data from over onehundred rivers and attempts to illustrate that particle size plays a role in determining under what discharge and slope a river will braid. Ferguson (1984) manipulates Parker's (1976) stability criterion and his generalized hydraulic geometry function (Parker, 1976) to produce an equation to discriminate the slope-discharge combination needed to cause braiding in a channel of a particular grain size. This is useful because it indicates an implied particle size dependency. The gently sloping lines in Figure 5.14 express the equation. Ferguson then used these data to develop a similar expression which more closely fits the data:

$$S = 0.042Q^{-0.49} D_{50}{}^{0.09} \tag{5.17}$$

where S is slope, Q is discharge in m^3/s and D_{50} is the median grain size in mm. By comparison, the original Leopold and Wolman (1957) curve is shown, illustrating that for gravel it tends to set the boundary too low. The Parker regime diagram, which uses Froude number slope, and depth to width ratios as axes is also being reevaluated. For example, Blondeau and Seminara (1984) reformulated the stability analysis by Parker with a different assumption about the effect of transverse bedslope on particle motion and suggest a modification of the Parker stability boundaries.

5.5 CHANNEL CHANGES

5.5.1 *Bank erosion*

One of the most difficult problems in fluvial geomorphology yet to be solved is the construction of a physically-based model that predicts with reasonable accuracy the erosion of channel banks. It is a problem involving soil and fluid mechanics, weathering and growth of vegetation. The meander models reviewed simulate bank erosion by driving it with a near bed velocity; bank properties are represented with a change in an arbitrary constant. Although valuable models, their application require calibration to individual rivers and reveal very little about bank erosion controls. Thorne (1982, 1981, Thorne and Tovey 1981, Thorne and Lewin 1979) has probably been the most active worker in the field problem of quantifying mechanical properties of river banks and their resistance to erosion.

Thorne (1982) divides bank erosion processes into two categories: fluvial entrainment and subaerial/subaqueous weathering and weakening. Fluvial entrainment acts both directly by removing bank material and indirectly by scouring the channel bed at the base of the bank, which leads to increased height and slope of the bank and subsequent bank failure. The process of fluvial entrainment must be treated in a similar manner to the entrainment of bed material, with the notable additions: the bank angle is sufficiently steep to require a slope-dependent gravitational term in the entrainment criterion, the bank material is often cohesive, and flow structure causing scour at the bank is often non-uniform and difficult to determine. The entrainment of cohesive material is still largely an unresolved problem. Grissinger (1982) discusses the present state of knowledge and emphasizes the complex interaction between the factors which influence cohesion in a soil. These factors include the primary soil properties, test conditions, bulk strength and composite soil properties, and hyraulic properties. Current studies in northern Mississippi by the USDA Sedimentation Laboratory (Grissinger 1982) are being conducted in an intensely instrumented watershed. Measurements of rainfall, runoff, sediment concentration, soil water content, land use, vegetative cover, and climatic conditions are being integrated to investigate the mode of failure of various stratigraphic horizons under varying natural conditions. Preliminary results are presented by Grissinger (1982) and the complete results should be forthcoming.

Subaerial/subaqueous weakening of the bank material is caused by the moisture condition of the soil in a variety of ways: positive pore water pressure reduces the effective strength, wetting and drying leads to the development of interped fissures and desiccation cracks which weaken the soil structure, freezing of pore water reduces granular interlocking, and the movement of pore water through the soil can leach clay particles and decrease the cohesion of the soil (Thorne 1982). Bank failures are typical of those encountered in soil mechanics: rotational slip, plane slip, and cantilever failure, with the main peculiarities associated with the unique hydraulic conditions. Soil shear strength can be measured by various engineering methods and Thorne (1981) describes the use of a borehole shear strength tester. The tensile strength is not as commonly measured in engineering analysis, but Thorne et al. (1980) describe measurements using a modified compression tester. The influence of vegetation on soil strength cannot be overemphasized, since it affects all of factors described above as controlling soil strength. Thorne and Tovey (1981) discuss the common occurrence of composite river banks wherein a strong vegetation mat overlies relatively weak silt deposits. This leads to undercutting of the bank and toppling of the vegetative mat into the channel. Other conditions can be locally important, such as permafrost in the banks of Arctic rivers. Church and Miles (Thorne 1982) describe a number of unique features associated with these channel banks.

Thorne (1981) presents a good summary of field methods available for studying bank erosion rates. The measurement of bank erosion rates can be done through either on-site measurements or remote sensing. On-site measurement is based on creating a detailed map of the channel reach using a plane table or tape and compass and then installing erosion pins for actual measurement of bank erosion. Reinforcing bar is inserted horizontally into the bank at various heights, such that the amount of erosion can be determined at various locations along the bank. The nature of the erosion process must be considered when determining the location and size of the erosion pins so that the erosion process is not disturbed. Thorne recommends the use of spray painted markings on gravelly bank material since the erodibility would be enhanced by erosion pin disturbance. In large rivers, the size of individual bank collapse events is large enough that measurement must be made from monumented stakes well back from the edge of the bank. Other considerations must be made on an individual basis. The use of aerial photography and/or historical maps provides an independent means of monitoring bank erosion. This has the advantage of allowing for an increased time period and channel length of study, but has the disadvantage of providing fewer details of the processes of erosion. Ultimately, both methods should be used together to characterize accurately the bank erosion processes (e.g. Thorne and Lewin 1979).

5.5.2 *Channel migration*

Channel migration has been investigated through the interaction of bed topography and planform geometry (Hooke and Harvey 1983, Milne 1982, Lewin 1978); through a mathematical model relating bank erosion to near-bank fluid velocity (references given above, Begin 1981, Rohrer 1984); or through a description and classification of planform changes through time, based on map and aerial photography analysis (Hooke 1984, Hooke and Harvey 1983, Nanson and Hickin 1983, Brice 1984). The work by Nanson and Hickin is particularly interesting because they report data that suggest a systematic relationship between outer bank erosion rates in bends and an average radius of curvature to width ratio. As described above, their work has attracted considerable attention because it provides a quantitative link between channel form and erosion rate which can readily be used in channel migration simulations. Parker (1984), among others, points out, however, that their data cannot correctly be used to predict the variation in migration rate through a single bend caused by systematic curvature changes because the data represent a kind of whole bend average. For the purposes of modelling, such channel migration data would be more useful if the combination of radius of curvature and distance through a bend from the crossing could be related empirically to a local, corresponding erosion rate. Such data would be much more difficult to obtain. Importantly, Nanson and Hickin report the finding by Humphrey (1976) that two bends in the same reach of channel (Fig. 5.13B) had essentially the same planform but one (left-hand channel) did not migrate in 21 years whereas the other (right-hand channel with broad point bar) migrated an average of 5 m/year. In response to Humphrey's observations Nanson and Hickin studied the dendrochronology of trees on successive scroll bars and found that over a onehundred and twenty year period the migration rates of the two bends were essentially identical. They correctly point out that studies of channel migration with the intention of establishing planform, migration rate relationships may need to have a much larger time base than is normally available from successive aerial photographs. Nanson and Hickin are currently attempting to expand their empirical relationship to other rivers. Such work, particularly where bank properties and distance through the bend are quantified, is greatly needed.

5.6 LARGE RIVERS

Recently, there has been a growing interest in the very large rivers of the world. The interest stems from the importance of large rivers as a sediment and nutrient source for the oceans, the unique scale availability for the study of basic fluvial processes, and the tectonic history revealed by the stratigraphy, sedimentology, and geomorphology of large rivers and their deltas. The excitement caused by these studies will undoubtedly continue to grow as more logistical problems are overcome by each new project.

Milliman and Meade (1983) provide recent data of the world-wide delivery of river sediment to the oceans. They note that approximately half of the total is supplied by large rivers with annual sediment discharges greater than 15×10^6 t. The difficulties inherent in calculating sediment yields from large drainage basins are described by Meade (1982) in his study of the Atlantic drainage of the United States. The present day sediment yield is affected by both the decreased sediment supply from modern farming practice and removal from storage within the drainage of large sediment loads left from land clearing by early settlers in the region. The problem is further complicated by manmade reservoirs and sediment accumulation in estuaries and coastal swamps. A more detailed routing of sediment through the Rio Orinoco in South America has been accomplished by Meade et al. (1983a, b). By observing the timing of peak water and sediment discharges at various locations along the river they have demonstrated both a consistent relation near the mouth of peak sediment discharge occurring during the rising stage, and a complex set of relations in the upstream reaches as sediment and water contributions of various tributaries interact with different timing. This interaction of tributaries requires that a wide range of fluvial processes be understood in order to explain the unique behaviour of each large river. This represents a good example of the importance of studying basic fluvial processes rather than attempting to develop global empirical relations.

Chinese scientists have recently begun a systematic study of fluvial processes in the Yangtze River (Pengzhang et al. 1983). They have set up a network of stations along the length of the Yangtze and created unified standards for data collection. They are studying both the operation of basic fluvial processes on a large scale and the particular engineering problems associated with large rivers.

Direct measurement of the hydraulic and sediment characteristics of large rivers is very difficult logistically. Nordin et al. (1983) present a means of rapid measurement of water and suspended-sediment discharge from a free-moving boat. The technique consists of a microwave positioning device and the combined use of a continuous reading Ott-type current meter and a depth-integrating suspended-sediment sampler. Dunne et al. (1983) have investigated some hydraulic parameters on the Amazon River by means of vertical profiles of velocity and concurrent suspended sediment sampling. They used the shear stress calculated from the velocity profiles to back calculate for the water surface slope and to predict the bedload transport rate. Mertes and Dunne (1983) are working with these bedload transport calculations and satellite imagery in an attempt to relate morphology variations along the Amazon to the observed bedload transport rates.

The importance of large rivers to studies of stratigraphy, sedimentology, and tectonic history was noted by Potter (1978) and Audley-Charles et al. (1979). Potter pointed out that large rivers are usually associated with tectonically controlled troughs, exist for long periods of time, and provide important stratigraphic information concerning the tectonic history of both the land mass and its adjacent marine basin. Audley-Charles et al. described further tectonic controls on large rivers and emphasized the importance of large deltas in terms of both natural resources and tectonic history. Undoubtedly, as more detailed work on the geomorphology of these large rivers is accomplished the tectonic histories will begin to emerge more clearly as both a result and necessary condition of understanding the geomorphic history.

The application of geomorphology to problems of neotectonics has received increased attention recently, as demonstrated by the 1983 Penrose Conference on Tectonic Geomorphology. Schumm et al. (1982), Watson et al. (1984) and Burnett and Schumm (1983)

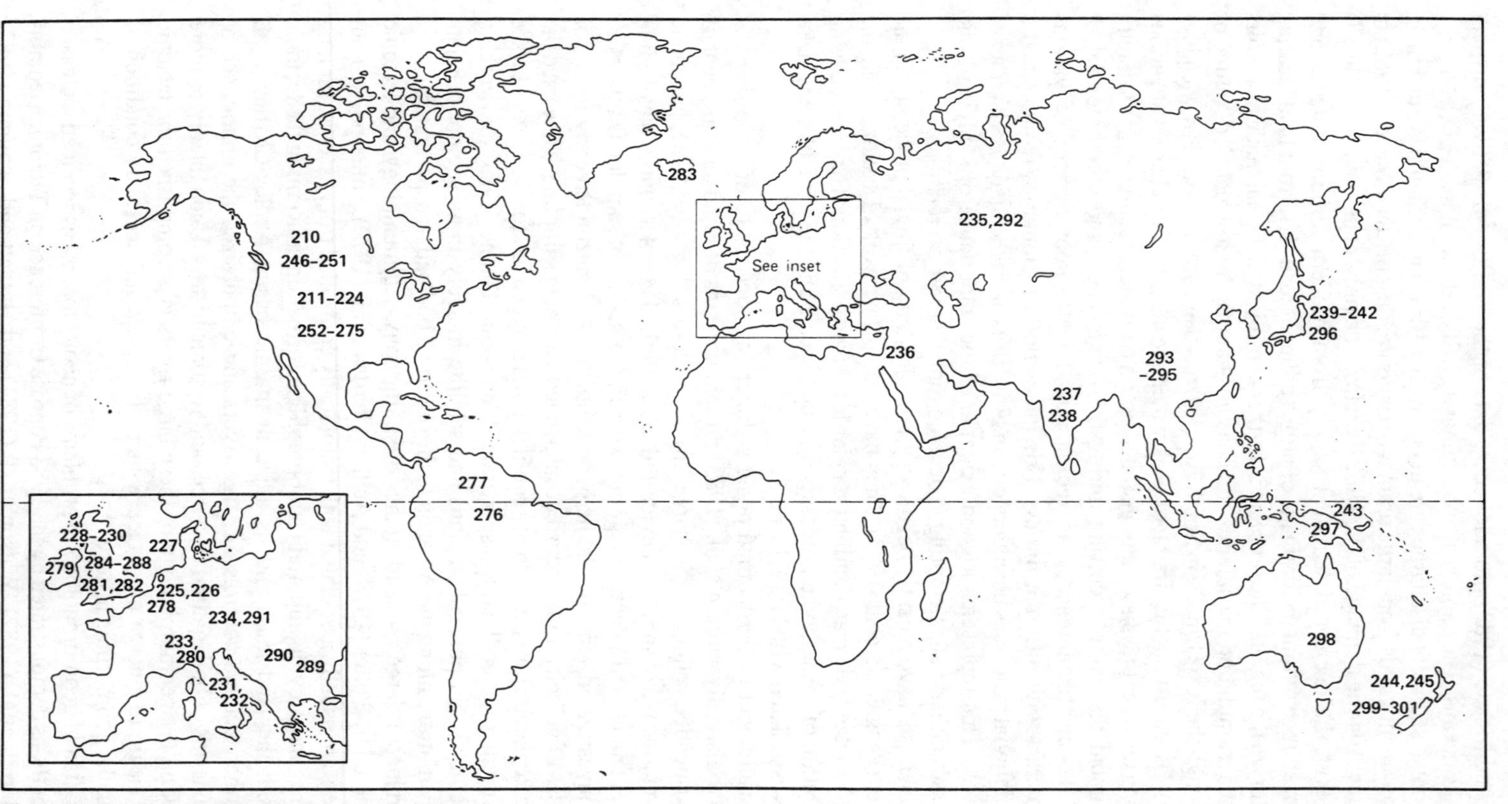

Figure 5.16. Location of fluvial experiments reported.

Table 5.1. Laboratory and theoretical studies of significance to fluvial field experiments.

Study No	Country	Researchers	Problem	Major objectives a	b	c	d	e	Sample reference
210	Canada	Ashmore, Parker	braiding				x	x	Ashmore (1982)
211	USA	Smith & McLean	flow through river bends	x					Smith & McLean (1984)
212	USA	Siegenthaler & Shen	flow through river bends	x				x	Siegenthaler & Shen (1984)
213	USA	Odgaard, Kennedy, Nokato	flow and bed morphology in bends	x	x	x	x	x	Kennedy et al. (1984)
214	USA-Japan	Parker, Ikeda, Sawai & co-workers	river meandering		x				Ikeda et al. (1981), Parker et al. (1982b), Parker (1984)
215	USA	Beck	river meandering		x		x		Beck (1984)
216	USA	Howard	river meandering				x		Howard (1984)
217	USA	Smith & McLean	form drag on dunes					x	Smith & McLean (1977)
218	USA	Brownlie	form drag on dunes					x	Brownlie (1983)
219	USA	Gust & Southard	flow resistance and sediment transport					x	Gust & Southard (1983)
220	USA	Bridge & Dominic	flow resistance and sediment transport					x	Bridge & Dominic (1984)
221	USA	Mantz	initial motion					x	Mantz (1983)
222	USA	Parker et al.	heteorogeneous bedload transport	x				x	Parker et al. (1982)
223	USA	Parker	channel geometry		x				Parker (1978a, b, 1979)
224	USA	Osterkamp	channel geometry				x		Osterkamp et al. (1983)
225	Netherlands	De Vriend, Kalkwijk, Struiksma	flow and bed morphology in bends	x	x		x		De Vriend & Struiksma (1984)
226	Netherlands	Olesen	alternate bars		x				Olesen (1984)
227	Denmark	Engelund & Fredsoe	form drag on dunes					x	Engelund & Fredsoe (1982)
228	UK	White & Day	bedload transport	x				x	White & Day (1982)
229	UK-USA	Bridge	river bends		x				Bridge (1984)
230	UK	Ferguson	river meandering				x		Ferguson (1984)
231	Italy	Becchi	flow resistance					x	Becchi (1983)
232	Italy	Blondeaux & Seminare	meandering and braiding		x				Blondeaux & Seminare (1984)
233	Switzerland	Jaeggi	alternate bars				x		Jaeggi (1984)
234	Poland	Gladki	flow resistance					x	Gladki (1979)
235	USSR	Nikitin, Deineka	suspended sediment	x					Nikitin & Deineka (1983)
236	Israel	Begin	river meandering				x		Begin (1981)
237	India	David & Gangadharaiah	initial motion					x	David & Gangaharaiah (1983)

Table 5.1 (continued).

Study No.	Country	Researchers	Problem	Major objectives a	b	c	d	e	Sample reference
238	India	Misri et al.	heterogeneous bedload	x				x	Misri et al. (1983)
239	Japan	Hasegama & Hamaoka	meandering	x	x	x			Yamaoka & Hasegama (1984)
240	Japan	H. Ikeda	bed features & sediment transport				x	x	H. Ikeda (1983)
241	Japan	S. Ikeda	alternate bars		x		x	x	S. Ikeda (1982, 1983)
242	Japan	Nakagawa, Hotsuta	alternate bars					x	Nakagawa & Hotsuta (1983)
243	Papua New Guinea	Pickup et al.	bedload waves		x				Pickup et al. (1983)
244	New Zealand	Weir	bedload waves	x					Weir (1983)
245	New Zealand	Davies & Tinker	river meandering				x		Davies & Tinker (1984)

a. Mathematical model to predict flow and boundary shear stress,
b. Mathematical model to predict channel morphology,
c. Mapping of flow, boundary shear stress and sediment transport fields in flumes,
d. Empirical analysis of channel form and behaviour,
e. Empirical analysis of friction, transport rate or grain size distribution.

have presented a thorough report on one such application. Zones of active uplift around the Mississippi River were investigated in order to determine if the fluvial response to the uplift is of sufficient magnitude to warrant consideration in engineering activities. They first used evidence from historical earthquake activity, deformed terrace surfaces, neck-cutoff frequency, and valley profile convexity to delineate the location and timing of recent uplift and demonstrate that it is still active. They then investigated the response of the river to the uplift by examining variations in the channel pattern associated with the convex valley profile. There may be a need for a more thorough clarification of what channel characteristics can be used as evidence for neotectonics (e.g. valley slope) and which characteristics should be intepreted in terms of activity thus delineated (e.g. sinuosity).

Another common application of fluvial geomorphology is in the evaluation of major engineering works on big rivers. Moreover, it is this context in which our difficulty in developing consistent predictive relations becomes obvious. Williams and Wolman (in press) have collected and analyzed data on the downstream effect of dams on alluvial rivers from the central United States. They show that over time periods of ten to thirty years the rivers show a wide variety of behaviour within a generally consistent response of degradation and channel widening downstream of the dam. Their observations also show that in many cases aggradation occurred downstream of the degradational reaches, presumably due to sediment input from tributaries. Some empirical relations were developed for the degradation and channel widening over time. These relations cannot be used for predictive purposes, since coefficients must be set for each river. However, the coefficients so calculated provide a means of parameterizing the channel response and may eventually be used to relate channel response to parameters such as sediment size,

Table 5.2. Fluvial process field experiments.

No.	Country	Researchers	Location	Problem	Major objectives a	b	c	d	e	Sample reference
246	Canada	Bray	Canadian rivers	flow resistance					x	Bray (1980, 1982)
247	Canada	Parker & Peterson	Canadian rivers	flow resistance					x	Parker & Peterson (1980)
248	Canada	Church, Jones	British Columbia	sediment transport periodicity		x				Church (1983), Church & Jones (1982)
249	Canada	Carson & La Pointe	North American rivers	meander planform		x				Carson & La Pointe (1983)
250	Canada	Nanson & Hickin	Beatton river	meander migration		x				Nanson & Hickin (1983)
251	Canada	Smith & Smith	Alberta, Saskatchewan	braiding, anastomosing	x	x		x	x	Smith & Smith (1980, 1984)
252	USA	Dietrich, Smith & Dunne	Muddy Creek, Wyoming	river meander mechanics	x	x				Dietrich & Smith (1983, 1984), Dietrich et al. (1979, 1984)
253	USA	Thorn, Rais	Fall River, Colo.	river meander	x	x	x			Thorn & Rais (1984)
254	USA	Chiu, Nordin, Hu	East Fork River, Wyoming	flow in bends			x			Chiu et al. (1984)
255	USA	Leopold	Western US rivers	controls on river planform			x			Leopold (1982)
256	USA	Bridge, Smith	Platte River	braiding mechanics	x	x	x	x		see J. Bridge, SUNY, Binghamton
257	USA	Prestegaard	Western US rivers	briading mechanics		x	x		x	Prestgaard (1983a, b)
258	USA	Smith & McLean	Columbia River, Washington	flow over dunes			x			Smith & McLean (1977a, b), McLean & Smith (1979)
259	USA	Dietrich & Smith	Muddy Creek, Wyo.	flow over dunes			x			Dietrich et al. (1984)
260	USA	Lisle	N.Calif. rivers	aggradation		x				Lisle (1982)
261	USA	Rohrer	Minnesota rivers	meanders			x			Rohrer (1984)
262	USA	Andrews	Western US rivers	initial motion					x	Andrews (1983, 1984)
263	USA	Leopold, Emmett, Meade, Nordin	East Fork River Wyoming	sediment transport periodicity		x				Meade et al. (1981), Emmett et al. (1983)
264	USA	Kelsey	Northern Calif.	hillslope-river			x			Kelsey (1980)
265	USA	Andrews	East Fork River	scour and fill		x	x			Andrews (1979a, b)
266	USA	McDonald, Keller	Northern Calif.	debris jams			x			Mosley (1981)
267	USA	Jackson & Beschta	Oregon Coast Range	sediment transport	x	x				Jackson & Beschta (1982)
268	USA	Lisle & Kelsey	Northern Calif.	obstruction controlled morphology		x			x	Lisle & Kelsey (1982, 1983)
269	USA	Andrews	Wyoming & Colorado	channel geometry		x				Andrews (1982, 1984)
270	USA	Brice	US rivers	channel planform		x				Brice (1984)

Table 5.2 (continued).

No.	Country	Researchers	Location	Problem	Major objectives a	b	c	d	e	Sample reference
271	USA	Blodgett & Stanley	Platte River	briading		x				Blodgett & Stanley (1980)
272	USA	Crowley	Platte River	briading		x				Crowley (1983)
273	USA	Grissinger	N. Mississippi	bank erosion		x				Grissinger (1982)
274	USA	Schumm, Watson, Harvey, Burnett	Mississippi River	tectonics & form		x				Schumm et al. (1982), Watson et al. (1984)
275	USA	Williams & Wolman	Central US rivers	downstream effects		x			x	Williams & Wolman (1984)
276	USA	Meade, Nordin, Dunne, Mertz	Amazon River	sediment transport	x	x			x	Meade et al. (1979, 1983a)
277	USA	Meade et al.	Orinoco River	sediment transport	x	x			x	Meade et al. (1983a, b).
278	Netherlands	De Vriend & Geldof	River Dammel	flow in bends			x			De Vriend & Geldof (1983)
279	UK	Bathurst, Thorn, Hay	English rivers	flow in bends			x		x	Bathurst et al. (1979)
280	UK	Gomez	Switzerland	armoring	x				x	Gomez (1983)
281	UK	Carling	British Rivers	initial motion					x	Carling (1983)
282	UK	Hammond et al.	tidal channel	initial motion					x	Hammond et al. (1984)
283	UK	Bluck	Welsh, Scottish, Icelandic rivers	grain size in bars				x		Bluck (1982)
284	UK	Lewin	Welsh rivers	alternate bars		x		x	x	Lewin (1976)
285	UK	Ferguson & Werritty	River Reshie, Scotland	gravel river behaviour, meandering		x				Ferguson & Werritty (1983), Ferguson (1984b)
286	UK	Thorne	Br. & Amer. rivers	bank erosion		x				Thorne (1982)
287	UK	Bridge & Jarvis	South Esk River	river meander mechanics	x		x	x		Bridge & Jarvis (1982)
288	UK	Hooke & Harvey	River Dane	river meander mechanics		x	x			Hooke & Harvey (1982)
289	Romania	Ichim et al.	Romanian rivers	response to flood flow	x				x	Ichim et al. (1980)
290	Hungary	Rakoczi	Danube river	river meanders	x	x	x			Rakoczi (1981)
291	Poland	Rachocki	Radunia rivere	sediment transport variation	x				x	Rachocki (1981)
292	USSR	Gergov	Russian streams	steep river bed morphology		x				Gergov (1983)

293	China	Guanghua, Pengzhang et al.	Yangtze River	sediment transport	x		x			Guanghua et al. (1983)
294	China	Ying	Minjang & Yangtze rivers	sediment transport	x				x	Ying (1983)
295	China	Zhengying	Chinese rivers	hyperconcentrated flows	x					Zhengying (1983)
296	Japan	Ashida et al.	Japanese Alps	mountain stream transport	x	x				Ashida et al. (1981)
297	Papua New Guinea	Pickup et al.	Fly River	bedload wave		x				Pickup et al. (1983)
298	Australia	Rust	Copper Creek	anastomosing		x				Rust (1981)
299	New Zealand	Pearce & Watson	NZ rivers	aggradatioin		x				Pearce & Watson (1983)
300	New Zealand	Mosley	NZ rivers	debris jams		x				Mosley (1981)
301	New Zealand	Whittaker & Jaeggi	NZ rivers	steep river bed morphology		x			x	Whittaker & Jaeggi (1982)

a. Monitoring of fluxes,
b. Mapping of channel form and behaviour,
c. Intensive mapping of sediment transport, flow, and/or boundary shear stress fields,
d. Sedimentology of deposits,
e. Analysis of friction and/or sorting.

sediment load, and peak discharge. Equally important is the number of unique responses observed, which once again require an interpretation based on an understanding of basic fluvial processes rather than grand empirical relations.

5.7 CONCLUSION

The interest in field experiments in fluvial geomorphology is truly an international one (Fig. 5.16 and Table 5.1). Frequent international meetings are providing a forum for rapid dissemination of ideas. It may well be that our understanding of rivers will be put to the greatest test in Asia and South America where massive hydroelectric and flood control projects are underway.

As in any field, important concepts in fluvial geomorphology long held as inviolate are now being reassessed in the light of recent theoretical and field studies. Emphasis is being placed on developing mechanistic theories, based as firmly as possible in fluid and sediment transport mechanics, that can be used to make numerical models of channel form and behaviour. Field studies are propelled by theoretical predictions, while theory is evolving with new detailed field observations. Computer-based numerical models are becoming commonplace and are allowing geomorphologists to explore the behaviour and form of rivers over large time and spatial scales. We can anticipate that future models will tend to grow in sophistication as intensive field studies reveal and quantify the fundamental processes controlling the interaction of flow, channel topography and sediment transport.

REFERENCES

Abrahams, A.D., 1984. 'Channel networks: a geomorphological perspective', *Water Resources Research*, 20, 161-188.

Allen, J.R.L., 1982. *Sedimentation structures: their character and physical basis*, 1 and 2, Elsevier Scientific Publishing Co.

Andrews, E.D., 1979a. Scour and fill in a stream channel, East Fork River, Western Wyoming, *US Geol. Surv.*, Prof. Paper 1117.

Andrews, E.D., 1979b. Hydraulic adjustment of the East Fork River, Wyoming, to the supply of sediment. In D.D. Rhodes and G.P. Williams (eds.), *Proc. tenth annual geomorphology symp.*, Binghamton, N.Y., 69-94.

Andrews, E.D., 1982. Bank stability and channel width adjustment, East Fork River, Wyoming, *Water Resources Research*, 18, 1184-1192.

Andrews, E.D., 1983. Entrainment of gravel from naturally sorted riverbed material, *Geol. Soc. Amer. Bull.*, 94, 1225-1231.

Andrews, E.D., 1984. Bed-material entrainment and hydraulic geometry of gravel-bed rivers in Colorado, *Geol. Soc. Amer. Bull.*, 95, 371-378.

Ashida, K. and Michiue, M., 1971. An investigation of river bed degradation downstream of a dam, *Proc. 14, congress of IAHR*, 3.

Ashida, K., Takahashi, T. and Sawada, T., 1981. Processes of sediment transport in mountain stream channels, *IAHS Pub.*, 132, 166-178.

Ashmore, P.E., 1982. Laboratory modelling of gravel braided stream morphology, *Earth Surface Processes and Landforms*, 7, 201-225.

Ashmore, P. and Parker, G., 1983. Confluence scour in coarse braided streams, *Water Resources Research*, 19, 392-402.

Audley-Charles, M.G., Curray, J.R. and Evans, G., 1979. Significance and origin of big rivers: a discussion. *J. Geol.*, 87, 122-123.

Bathurst, J.C., Li, R.M. and Simons, D.B., 1981. Resistance equation for large-scale roughness, *J. Hydraulics Div., Amer. Soc. Civil Engineers*, 107, no. HY12, Proc. Pap. 16743, 1593-1613.

Bathurst, J.C., Thorne, C.R. and Hey, R.D., 1979. Secondary flow and shear stress at river bends, *J. Hydraulics Div., Amer. Soc. Civil Engineers*, 105, 1277-1295.

Becchi, I., 1983. Effects of sediment motion on river roughness, *Proc. of second international symp. on river sedimentation*, Water Resources and Electric Power Press, China, 162-172.

Beck, S., 1984. Mathematical modelling of meander interactions. In C. Elliot (ed.), *River meandering*, Amer. Soc. Civil Engineers, 932-941.

Beck, S., Melfi, D.A. and Yalamanchili, K., 1984. Lateral migration of the Genesee River. In C. Elliot (ed.), *River meandering*, Amer. Soc. Civil Engineers, 510-517.

Begin, Z.B., 1981. Stream curvature and bank erosion: a model based on the momentum equation, *J. Geol.*, 89, 497-504.

Bjorn, T.C. et al. 1977. Transport of granitic sediment in streams and its effects on insects and fish, *Forest, Wildlife and Range Experiment Station*, Bull. 17, Univ. of Idaho.

Blodgett, R.H. and Stanley, K.O., 1980. Stratification, bedforms, and discharge relations of the Platte braided river system, Nebraska, *J. Sedimentary Petrology*, 50, 139-148.

Blondeaux, P. and Seinarci, G., 1984. Bed topography and instabilities in sinuous channels. In C. Elliot (ed.), *River meandering*, Amer. Soc. Civil Engineers, 747-758.

Bluck, B.J., 1982. Texture of gravel bars in braided streams. In R.D. Hey, J.C. Bathurst and C.B. Thorne (eds.), *Gravel-bed rivers*, John Wiley and Sons, Ltd., Chichester, 339-355.

Bray, D.I., 1980. Evaluation of effective boundary roughness for gravel-bed rivers, *Canadian J. Civil Engineers*, 7, 392-397.

Bray, D.I., 1982. Flow resistance in gravel-bed rivers. In R.D. Hey, J.C. Bathurst and C.R. Thorne (eds.), *Gravel-bed rivers*, J. Wiley and Sons, Ltd., 109-137.

Brice, J.C., 1973. Meandering pattern of the White River in Indiana – an analysis. In M. Morisawa (ed.), *Fluvial geomorphology*, SUNY Binghamton, Publ. in *Geomorph.*, 179-200.

Brice, J.C., 1984. Planform properties of meandering rivers. In C. Elliot (ed.), *River meandering*, Amer. Soc. Civil Engineers, 1-15.

Bridge, J.S., 1984. Flow and sedimentary processes in river beds: comparisons of field observations and theory. In C. Elliot (ed.), *River meandering*, Amer. Soc. Civil Engineers 857-872.

Bridge, J.S. and Dominic, C.F., 1984. Bed-load grain velocities and sediment transport rates, *Water Resources Research* 20, 476-490.

Bridge, J.S. and Jarvis, J., 1982. The dynamics of a river bend: a study of flow and sedimentary processes, *Sedimentology* 29, 499-541.

Brownlie, W.R., 1983. Flow depth in sand-bed channels, *J. Hydraul. Div., Amer. Soc. Civil Engineers* 109, 959-990.

Burnett, A.W. and Schumm, S.A., 1983. Alluvial river response to neotectonic deformation in Louisiana and Mississippi, *Science* 223, 49-50.

Carling, P.A., 1983. Threshold of coarse sediment transport in broad and narrow natural streams, *Earth Surface Processes and Landforms* 8, 1-18.

Carson, M.A. and La Pointe, M.F., 1983. The inherent asymmetry of river meander planform, *J. Geol.* 91, 41-55.

Cavazza, S., 1981. Experimental investigations on the initiation of bedload transport in gravel rivers. In *Erosion and sediment transport measurement*, IAHS Pubs. 133, 53-61.

Chang, H.H., 1979. Geometry of rivers in regime, *Amer. Soc. Civil Engineers, J. Hydraul. Div.* 105, no. HY6, 691-706.

Chang, H., Simons, D.B. and Woolhiser, D.A., 1971. Flume experiments on alternate bar formation, *Waterways, Harbors, and Coastal Eng. Div., Amer. Soc. Civil Engineers*, WWI, 155-165.

Chiu, C., Nordin, C.F. and Hu, W.D., 1984. A method for mathematical modelling and computation of hydraulic processes in a river bed. In C. Elliot (ed.), *River meandering*, Amer. Soc. Civil Engineers, 829-842.

Church, M., 1982. Pattern of instability in a wandering gravel bed channel, *Special publication international assoc. sedimentology* 6, 169-180.

Church, M. and Jones, D., 1982. Channel bars in gravel-bed rivers. In R.D. Hey, J.C. Bathurst and C.B. Thorne (eds.), *Gravel-bed rivers*, John Wiley and Sons, Ltd., Chichester, 291-338.

Crowley, K.D., 1983. Large-scale bed configurations (macroforms), Platte River Basin, Colorado and Nebraska: Primary structures and formative processes, *Geol. Soc. Amer. Bull.* 94, 117-133.

David, K.J. and Gangadharaiah, T., 1983. The effect of nonuniformity in grain size on the initiation of grain motion *Proc. 2nd international symp. on river sedimentation*, Water Resources and Electric Power Press,China, 434-439.

Davies, T.R.H. and Pearce, A.J. (eds.), 1981. *Erosion and sediment transport in Pacific Rim steeplands*, IAHS Pub. 132.

Davies, T.R.H. and Tinker, C.C., 1984. Fundamental characteristics of stream meanders, *Geol. Soc. Amer. Bull.* 95, 505-512.

De Vriend, H.J. and Geldoff, H.J., 1983. Main flow velocity in short river bends, *J. Hydraul. Div., Amer. Soc. Civil Engineers* 109, 991-1011.

De Vriend, H.J. and Struiksma, N., 1984. Flow and bed deformation in river bends. In C. Elliot (ed.), *River meandering*, Amer. Soc. Civil Engineers, 810-828.

Dietrich, W.E., 1982. Settling velocity of natural particles, *Water Resources Research* 18, 1615-1626.

Dietrich, W.E. and Smith, J.D., 1983. Influence of the point bar on flow through curved channels, *Water Resources Research* 19, 1173-1192.

Dietrich, W.E. and Smith, J.D., 1984a. Bedload transport in a river meander, *Water Resoures Research* 20, 1355-1380.

Dietrich, W.E. and Smith, J.D., 1984b. Processes controlling the equilibrium bed morphology in river meanders. In C. Elliot (ed.), *River meandering*, Amer. Soc. Civil Engineers, 759-769.

Dietrich, W.E., Smith, J.D. and Dunne, T., 1979. Flow and sediment transport in a sand-bedded meander, *J. Geol.* 87, 305-315.

Dietrich, W.E., Smith, J.D. and Dunne, T., 1984. Boundary shear stress, sediment transport and bed morphology in a sand-bedded river meander during high and low flow. In C. Elliot (ed.), *River meandering*, Amer. Soc. Civil Engineers, 632-639.

Dunne, T., Mertes, L.A.K. and Meade, R.H., 1983. Hydraulics and sediment transport in the Amazon River system (abstr.), *EOS* 64, no. 45, 697.

Egiazaroff, I.V., 1965. Calculations of nonuniform sediment concentrations, *J. Hydraul. Div., Proc. Amer. Soc. Civil Engineers* HY-4.

Einstein, H.A. and Shen, H.W., 1964. A study of meandering in straight alluvial channels, *J. Geophysical Research* 69, 5239-5247.

Elliott, C.M. and Pokrefke, T.J., 1964. Channel stabilization in a straight river reach. In C. Elliot (ed.), *River meandering*, Amer. Soc. Civil Engineers, 873-884.

Emmett, W.W., Leopold, L.B. and Myricks, R.M., 1983. Some characteristics of fluvial processes in rivers, *Proc. 2nd international symp. on river sedimentation,* Water Resources and Electric Power Press, China, 730-754.

Engelund, F., 1974. Flow and bed topography in channel bends, *J. Hydraul. Div., Amer. Soc. Civil Engineers* 100 (HY11), 1631-1648.

Engelund, F., 1975. Instability of flow in curved alluvial channel, *J. Fluid Mechanics* 72, 145-160.

Engelund, F. and Fredsoe, J., 1982. Sediment ripples and dunes, *Annual Review of Fluid Mechanics* 14, 13-37.

Engelund, F. and Skovgaard, O., 1973. On the origin of meandering and braiding in alluvial streams, *J. Fluid Mechanics* 57, pt. 2, p. 289.

Fenton, J.D. and Abbott, J.E., 1977. Initial movement of grains in a stream bed: the effects of relative protrusion, *Proc. Royal Soc. London*, pt. A 352, 532-537.

Ferguson, R.I., 1973. Regular meander path models, *Water Resources Research* 9, 1079-1086.

Ferguson, R.I., 1976. Disturbed periodic model for river meanders, *Earth Surface Processes* 1, 337-347.

Ferguson, R.I., 1984. Kinematic model of meander migration. In C. Elliot (ed.), *River meandering,* Amer. Soc. Civil Engineers, 942-951.

Ferguson, R.I., 1984b. The threshold between meandering and braiding, in Channels and Channel

Control Structures. In K.V.H. Smith (ed.), *Proc. 1st international conf. hydraulic design in water res. eng.*, Computation Mech. Center Pub., Springer Verlag, N.Y., 6-15 to 6-20.

Ferguson, R.I. and Werritty A., 1983. Bar development and channel changes in the gravelly River Reshie, Scotland, *Special publication international assoc. sedimentology* 6, 181-193.

Fredsoe, J., 1978. Meandering and braiding of rivers, *J. Fluid Mechanics*. 84, pt. 4, p. 609.

Gergov, G., 1983. Peculiarities of the mountain river forms, *Proc. 2nd international symp. on river sedimentation*, Water Resources and Electric Power Press, China, 682-691.

Gladki, H., 1979. Resistance to flow in alluvial channels with coarse bed material, *J. Hydraulics Research* 17, 121-128.

Gomez, B., 1983. Temporal variations in bedload transport rates: the effect of progressive armouring, *Earth Surface Processes and Landforms*, 8, 41-54.

Grissinger, E.H., 1982. Bank erosion of cohesive materials. In R.D. Hey, J.C. Bathurst and C.B. Thorn (eds.), *Gravel-bed rivers,* John Wiley and Sons, Ltd., Chichester, 273-287.

Guanghua, H., Huanjin, G. and Yuchen, W., 1983. Study on sampling techniques of bedload in Yangtze River, *Proc. 2nd international symp. on river sedimentation*, Water Resources and Electric Power Press, China, 997-1016.

Gust, G. and Southard, J.B., 1983. Effects of weak bedload on the universal laws of the wall, *J. Geophysical Research* 88, 5939-5952.

Gustavson, T.C., 1978. Bedforms and stratification types of modern gravel meander lobes, Nueces River Texas, *Sedimentology* 25, 401-426.

Hammond, F.D.C., Heathershaw, A.D. and Langhorne, D.N., 1984. A comparison between Shields' threshold criterion and the movement of loosely packed gravel in a tidal channel, *Sedimentology* 31, 51-62.

Hayashi, T. and Ozaki, S., 1980a. Alluvial bedform analysis I: formation of alternating bars and braids. In H.W. Shen and H. Kikkawa (eds.), *Application of stochastic processes in sediment transport*, Water Resource Publications, Littleform, Colorado, USA.

Hayashi, T. and Ozaki, S., 1980b. On the saltation heights and step lengths of sediment particles in the bedload layer, *International symp. on river sedimentation*, Beijing, China.

Hayashi, T., Ozaki, S. and Ichibashi, T., 1980. Study on bedload transport of sediment mixture, *Proc. 24th Japanese conference on hydraulics*.

Hey, R.D., 1982. Design equations for mobile gravel-bed rivers. In R.D. Hey, J.C. Bathurst and C.B. Thorne (eds.), *Gravel-bed rivers*, Wiley and Sons, 553-580.

Hey, R.D., Bathurst, J.C. and Thorne, C.R., 1982. *Gravel-bed rivers*, Wiley-Interscience Pub., Wiley and Sons.

Hooke, J.M. and Harvey, A.M., 1983. Meander changes in relation to bend morphology and secondary flows, *International Assoc. Sedimentology special publication* 6, 121-132.

Hooke, R. LeB., 1975. Distribution of sediment transport and shear stress in a meander bend, *J. Geol.* 83, 543-565.

Howard, A., 1984. Simulation model of meandering. In C. Elliot (ed.), *River meandering*, Amer. Soc. Civil Engineers, 952-963.

Humphrey, N.F., 1978. An aerial photographic study of the meander behaviour of alluvial rivers in Northeast British Columbia and Northern Alberta. Unpublished B.Sc. Thesis Dept. Geography, Univ. British Columbia, Vancouver.

Ichim, I., Batuca, D. and Radoane, M., 1980. Problems of the dynamics of some Romanian river channels, *Paper presented at the meeting of the Commission of Field Experiments in Geomorphology*, 1980, Kyoto, Japan.

Ikeda, S., 1982. Incipient motion of sand particles on side slopes, *J. Hydraulics Div., Amer. Soc. Civil Engineers* 108, no. HY1, 95-114.

Ikeda, S., 1983. *Experiments on bedload transport, bed forms and sedimentary structures using fine gravel in the 4-meter-wide flume*, Env. Res. Center Paper No. 2, University of Tsukuba, Ibaraki, Japan.

Ikeda, S., 1984. Prediction of alternate bar wavelength and height, *J. Hydraulic Engineering* 110, 371-386.

Ikeda, S., Parker, G. and Sawai, K., 1981. Theory of river meanders: Part I, linear development, *J. Fluid Mechanics* 112, 363-377.

Ikeda, S., Parker, G. and Sawai, K., 1982. Theory of river meanders: Part 2, non-linear deformation of finite amplitude bends, *J. Fluid Mechanics* 115, 303-314.

Jackson, W.C. and Beschta, R.C., 1982. A model of two-phase bedload transport in an Oregon Coast Range stream, *Earth Surface Processes and Landforms* 7, 517-727.

Jaeggi, M.N.R., 1984. Formation and effects of alternate bars, *Amer. Soc. Civil Engineers, J. Hydraulic Engineering* 110, 142-156.

Jaeggi, M. and Smart, G., 1982. Discussion of channel bars in gravel-bed rivers. In R.D. Hey, J.C. Bathurst and C.R. Thorne (eds.), *Gravel-Bed rivers*, Wiley and Sons, p. 325.

Kalkwijk, J.P.Th. and De Vriend, H.J., 1980. Computations of the flow in shallow riverbends, *J. Hydraulic Research*, Delft, The Netherlands, 18, 327-342.

Keller, E.A. and Melhorn, W.N., 1978. Rhythmic spacing and origin of pools and riffles, *Geol. Soc. Amer. Bull.* 89, 723-730.

Keller, E.A. and Swanson, F.J., 1979. Effects of large organic material on channel form and fluvial processes, *Earth Surface Processes* 4, 356-380.

Keller, E.A. and Tally, T., 1979. Effects of large organic debris on channel form and fluvial processes in the Coastal Redwood Environment. In D.D. Rhodes and G.P. Williams (eds.), *Adjustments of the fluvial system*, Kendall/Hunt Pub. Co.

Kelsey, H.M., 1980. A sediment budget and analysis of geomorphic process in the Van Duzen River Basin, North Coastal California, 1941-1975, *Geol. Soc. Amer. Bull.* 91, 1191-1216.

Kennedy, J.F., Nokato, T. and Odgaard, A.J., 1984. Analysis, numerical modelling and experimental investigation of flow in riverbends. In C. Elliot (ed.), *River meandering*, Amer. Soc. Civil Engineers, 843-856.

Kinosita, R., 1961. *Study in the channel evolution of the Isi Kari river*, Bureau of Resonics, Dept. of Science and Tech., Japan (in Japanese).

Klingeman, P.C. and Emmett, W.W., 1982. Gravel bedload transport processes. In R.D. Hey, J.C. Bathurst and C.R. Thorne (eds.), *Gravel-bed rivers*, John Wiley and Sons, Ltd., 141-179.

Kondrat'yev, N. Ye., 1968. Hydromorphological principles of computations of free meandering, I. Signs and indexes of free meandering, *Soviet Hydrology, Transl.* 7, 309-335.

Lane, E.W., 1955. Design of stable channels, *Trans. Amer. Soc. Civil Engineers* 120, 1234-1279.

Langbein, W.B., 1964. Geometry of river channels, *J. Hydraulics Div., Amer. Soc. Civil Engineers* 90, 301-312.

Langbein, W.B. and Leopold, L.B., 1966. River meanders – theory of minimum variance, *US Geol. Surv. Prof. Paper 422 H*.

Leopold, L.B., 1982. Water surface topography in river channels and implications for meander development. In R.D. Hey, J.C. Bathurst & C.R. Thorne (eds.), *Gravel-Bed rivers*, John Wiley and Sons, Ltd., Chichester, 359-388.

Leopold, L.B. and Maddock, T. Jr., 1953. The hydraulic geometry of stream channels and some physiographic implications, *US Geol. Surv. Prof. Paper* 252.

Leopold, L.B. and Wolman, M.G., 1957. River channel patterns: braided, meandering, and straight, *US Geol. Surv. Prof. Paper* 232 B.

Lewin, J., 1976. Initiation of bedforms and meanders in coarse-grained sediment, *Geol. Soc. Amer. Bull* 87, 281-285.

Lewin, J., 1978. Meander development and floodplain sedimentation:a case study from mid-Wales, *Geol. J.* 13, 25-36.

Lisle, T.E., 1981. Recovery of aggraded stream channels at gauging stations in northern California and southern Oregon, *IAHS Pub.* 132, 189-211.

Lisle, T.E., 1982. Effects of aggradation and degradation on riffle-pool morphology in natural gravel channels, Northwestern California, *Water Resources Research* 18, 1643-1651.

Lisle, T.E. and Kelsey, H.M., 1982. Effects of large roughness elements on thalweg course and pool spacing, Jacoby Creek, northwestern California (abst.). In L.B. Leopold (ed.), *Amer. geomorph. field group guidebook*.

Lisle, T.E. and Kelsey, H.M., 1983. Thalweg attachment to large roughness elements, Jacoby Creek, northwestern California. *Abst. with Programs 1983, Geol. Soc. Amer.* 15, p. 628.

Mantz, P.A., 1983. Review of laboratory sediment transport research using fine sediments, *Proc. 2nd international symp. on river sedimentation*, Water Resources and Electric Power Press, China, 532-557.

McLean, S.R. and Smith, J.P., 1979. Turbulence measurements in the boundary layer over a sand wave field, *J. Geophysical Research* 84, 7791-7808.

Meade, R.H., 1982. Sources, sinks, and storage of river sediment in the Atlantic drainage of the United States, *J. Geol.* 90, 235-252.

Meade, R.H., Emmett, W.W. and Myrick, R.M., 1981. Movement and storage of bed material during 1979 in East Fork River, Wyoming, USA. *Erosion and sediment transport in pacific rim steeplands*, IAHS, Pub. no. 132, 225-235.

Meade, R.H., Nordin, C.F., Curtis, W.F., Costa Rodrigues, F.M., do Vale, C.M. and Edmond, J.M., 1979. Sediment loads in the Amazon River, *Nature* 278, 161-163.

Meade, R.H., Nordin, C.F. and Dunne, T., 1983. Movement and storage of suspended sediment in the Amazon and Orinoco Rivers (abstr.), *EOS* 64, 697.

Meade, R.H., Nordin, C.F., Hernandez, D.P., Mejia, A.B. and Godoy, J.M.P., 1983. Sediment and water discharge in Rio Orinoco, Venezuela and Colombia, *Proc. 2nd International Symp. on River Sedimentation*, Water Resources and Electric Power Press, Nanjing, China, 1134-1144.

Mertes, L.A.K. and Dunne, T., 1983. Channel changes on the Amazon River (abstr.), *EOS* 64, 697.

Milhous, R.T., 1973. Sediment transport in a gravel-bottomed stream. Unpublished Ph.D. thesis, Oregon State Univ. Corvallis.

Milhous, R.T. and Thorne, C.R., 1982. Discussion of gravel bedload transport processes. In R.D. Hey, J.C. Bathurst and C.R. Thorn (eds.), *Gravel-bed rivers*, Wiley and Sons, 173-175.

Milliman, J.D. and Meade, R.H., 1983. World-wide delivery of river sediment to the oceans, *J. Geol.* 91, 1-21.

Milne, J.A., 1982. Bedforms and bend-arc spacings of some coarse-bedload channels in Upland Britain, *Earth Surface Processes and Landforms* 7, 227-240.

Misri, R.L., Garde, R.J. and Rajn, K.G., 1983. Experiments on bedload transport of nonuniform sands and gravels, *Proc. 2nd international symp. on river sedimentation*, Water Resources and Electric Power Press, China, 440-450.

Mosley, M.P., 1981. The influence of organic debris on channel morphology and bedload transport in a New Zealand forest stream, *Earth Surface Processes and Landforms* 6, 571-579.

Nakagawa, T., 1983. Boundary effects on stream meandering and river morphology, *Sedimentology* 30, 117-127.

Nakagawa, T. and Hotsuta, M., 1984. Note on boundary effects on stream meandering, *Sedimentology*, 31, 119-122.

Nanson, G.C. and Hickin, E.J., 1983. Channel migration and incision on the Beatton River, *Amer. Soc. Civil Engineers, J. Hydraulic Engineering* 109, no. 3, 327-336.

Nikitin, I.K. and Deineka, V.I., 1983. The structure of the channel's suspended stream and its registration as while projecting the engineering hydroconstruction, *Proc. 2nd international symp. on river sedimentation*, Water Resources and Electric Power Press, China, 239-250.

Nordin, C.F., Cranston, C.C. and Mejia, A.B., 1983. New technology for measuring water and suspended-sediment discharge of large rivers, *Proc. 2nd international symp. on river sedimentation*, Water Resources and Electric Power Press, China, 1145-1158.

Olesen, K.W., 1984. Alternate bars on and meandering of alluvial river. In. C. Elliot (ed.), *River meandering*, Amer. Soc. Civil Engineers, 873-884.

Osterkamp, L., Lane, J. and Foster, G.R., 1983. An analytical treatment of channel-morphology relations, *US Geol. Surv. Prof. Paper 1288*.

Parker, G., 1976. On the cause and the characteristic scales of meandering and braiding in rivers, *J. Fluid Mechanics* 76, 457-480.

Parker, G., 1978a. Self-formed straight rivers with equilibrium banks and mobile bed, part I, The sand-silt river, *J. Fluid Mechanics* 89, 109-125.

Parker, G., 1978b. Self-formed straight rivers with equilibrium banks and mobile bed, part I, The sand-silt river, *J. Fluid Mechanics* 89, 127-146.

Parker, G., 1979. Hydraulic geometry of active gravel rivers, *J. Hydraulics Div., Amer. Soc. Civil Engineers*, HY9, 105, 1185-1201.

Parker, G., 1984. Theory of meander bend formations. In C. Elliot (ed.), *River meandering*, Amer. Soc. Civil Engineers, 727-733.

Parker, G. and Andres, D., 1976. Detrimental effect of river channelization, *Proc. Symp. on inland waters for navigation, flood control. and water diversion.*, Colo. State Univ., Rivers '76, 1248-1266.

Parker, G., Dhamotharan, S. and Stefan, H., 1982. Model experiments on mobile, paved gravel bed stream, *Water Resources Research* 18, 1395-1408.

Parker, G., Diplas, P. and Akiyama, J., 1983. Meander bends of high amplitude, *J. Hydraulic Engineering* 109, 1323-1337.

Parker, G., Higgins, R.J. and Grant, I., 1982. Modelling sediment transport as a moving wave – the transfer and deposition of mining waste, *J. Hydrology* 60, 281-301.

Parker, G. and Klingeman, P.C., 1982. On why gravel bed streams are paved, *Water Resources Research* 18, 1409-1423.

Parker, G. and Peterson, A.W., 1980. Bar resistance of gravel-bed streams, *J. Hydraulic Div., Amer. Soc. Civil Engineers* 106, 1559-1575.

Parker, G., Sawai, K. and Ikeda, S., 1982. Bend theory of river meanders, pt. 2, nonlinear deformation of finite-amplitude bends, *J. Fluid Mechanics* 115, 303-314.

Pearce, A.J. and Watson, A., 1983. Medium term effects of two landsliding episodes on channel storage of sediment, *Earth Surface Processes and Landforms*, 8, 29-39.

Pengzhang, Z., Jiazhong, Z., Yinglong, Z. and Wenxiang, G., 1983. Measurement of sediment and study on the fluvial process in Yangtze River (Engl. abstr.), *Proc. 2nd international symp. on river sedimentation*, Water Resources and Electric Power Press, China.

Pickup, G., Higgins, R.J., and Grant, I., 1983. Modelling sediment transport as a moving wave – the transfer and deposition of mining waste, *J. Hydrology* 60, 281-301.

Potter, P.E., 1978. Significance and origin of big rivers, *J. Geol.* 86, 13-33.

Power, M.E. and Mathews, W.J., 1983. Algae-grazing minonous (composition anomalum), piscivorous bass (micropterus spp.) and the distribution of attached algae in a small prairie-margin stream, *Ecologia* 60, 328-332.

Prestgaard, K.L., 1983a. Bar resistance in gravel bed streams at bankfull stage, *Water Resources Research* 19, 472-476.

Prestgaard, K.L., 1983b. Variables influencing water-surface slopes in gravel-bed streams at bankfull stage, *Geol. Soc. Amer. Bull.* 94, 673-678.

Rachocki, A., 1981. Forms of sand-gravel streaks in contemporary river channels, *Questiones Geographical* 7, 91-96.

Rakoczi, L., 1983. Movement of gravel bed load in sharp river bends, *Proc. 2nd international symp. on river sedimentation*, Water Resources and Electric Power Press, China, 1084-1097.

Raudkivi, A.J., 1976. *Loose boundary hydraulics*, 2nd Ed. Pergamon Press.

Richards, K., 1982. *Rivers: form and process in alluvial channels*, Methuen and Co., Ltd., NY.

Rohrer, W.L.M., 1984. Effects of flow and bank material on meander migration in alluvial rivers. In C. Elliot (ed.), *River meandering*, Amer. Soc. Civil Engineers, 770-780.

Rust, B.R., 1981. Sedimentation in an arid-zone anastomosing fluvial system: Cooper's Creek, Central Australia, *J. Sedimentary Petrology* 51, 745-755.

Saucier, R.T., 1984. Historic changes in current river meander regime. In C. Elliot (ed.), *River meandering*, Amer. Soc. Civil Engineers, 180-190.

Schumm, S.A., 1977. *The fluvial system*, Wiley-Interscience Pub.

Schumm, S.A., 1981. Evolution and response of the fluvial system, sedimentologic implications, *Soc. Economic Paleontologists and Mineralogists* Special Publication 31, 19-29.

Schumm, S.A. and Parker, R.S., 1973. Implications of complex response of drainage systems for quaternary alluvial stratigraphy, *Nat. Phys. Sci.* 243, 99-100.

Schumm, S.A., Watson, C.C. and Burnett, A.W., 1982. *Investigation of neotectonic activity within the lower Mississippi valley division, phase 1*, US Army Corps of Engineers, Potamology Prog. rep. 2, Vicksburg, MI.

Shen, H.W. and Komura, S., 1968. Meandering tendencies in straight alluvial channels, *J. Hydraulics. Div., Amer. Soc. Civil Engineers* 94,997-1046.

Shen, H.W. and Lu, J., 1983. Development and prediction of bed armouring, *J. Hydraulics. Div., Amer. Soc. Civil Engineers* 109, 611-629.

Siegenthaler, M.C., 1984. Analysis of selected models of velocity and bed shear stress distribution in meanders. Unpublished Masters thesis, Colorado State University.

Siegenthaler, M.C. and Shen, H.W., 1984. Shear stress uncertainties in bends from equations. In C. Elliot (ed.), *River meandering*, Amer. Soc. Civil Engineers, 662-674.

Simons, D.B. and Richardson, E.V., 1966. Resistance to flow in alluvial channels, *US Geol. Surv. Prof. Paper 422J*.

Smith, D.G. and Smith, N.D., 1980. Sedimentation in anastomosed river systems: examples from alluvial valleys near Banff, Alberta, *J. Sedimentary Petrology* 50, 157-164.

Smith, J.D. and McLean, S.R., 1977a. Spatially averaged flow over a wavy boundary, *J. Geophysical Research* 82, 1735-1747.

Smith, J.D. and McLean, S.R., 1977b. Boundary layer adjustments to bottom topography and suspended sediment. In J.C.J. Nihoul (ed.), *Bottom turbulence*, Elsevier Oceanographic Series, vol. 19, 123-151.

Smith, J.D. and McLean, S.R., 1984. A model for meandering streams, *Water Resources Research*, 20, 1301-1315.

Smith, N.D., 1978. Some comments on terminology for bars in shallow rivers. In A.D. Miall (ed.), *Fluvial sedimentology*, Canadian Soc. of Petroleum Geologists, Calgary, Canada, 85-88.

Smith, N.D., Southard, J.B. and Kang, S.D., 1981. Gravel transport in a braided glacial stream, *abstracts to the Keele conference, modern and ancient fluvial systems*, International Assoc. of Sedimentology.

Smith, N.D. and Smith, D.G., 1984. William River: an outstanding example of channel widening and braiding caused by bedload addition, *Geol.* 12, 78-82.

Song, C.C.S. and Yang, C.T., 1980. Minimum stream power: theory, Amer. Soc. Civil Engineers, *J. Hydraulics Div.* 106, no. HY9, 1477-1487.

Southard, J.B., 1981. Field and laboratory studies of braiding in shallow gravel-bed streams (abst.), *Keele conference, modern and ancient fluvial systems*, International Assoc. Sedimentology.

Stelczer, K., 1981. *Bed-load transport: theory and practice*, Water Res. Pubs, Colorado, USA, 295 p.

Thorne, C.R., 1982. Processes and mechanisms of river bank erosion. In R.D. Hey, J.C. Bathurst and C.B. Thorn (eds.), *Gravel-bed rivers*, John Wiley and Sons, Ltd., Chichester, 227-271.

Thorne, C.R., 1981. Field measurements of bank erosion and bank material strength. *Erosion and sediment transport measurement*, IAHS Pub. 133, 503-512.

Thorne, C.R. and Lewin, J., 1979. Bank processes, bed material movement and planform development in a meandering river. In R.D. Rhodes and G.P. Williams (eds.), *Adjustments of the fluvial system*, Kendall/Hunt, Dubuque.

Thorne, C.R. and Rais, S., 1984. Secondary current measurements in a meandering river. In C. Elliot (ed.), *River meandering*, Amer. Soc. Civil Engineers, 675-686.

Thorne, C.R. and Tovey, N.K., 1981. Stability of composite river banks, *Earth Surface Processes and Landforms* 6, 469-484.

Thorne, C.R., Tovey, N.K. and Bryant, R., 1980. Recording unconfined tension tester, *J. Geotech. Engineering Div., Amer. Soc. Civil Engineers* 106 (GT11), 1269-1273.

Vanoni, V.A, 1964. *Measurements of critical shear stress*, California Institute of Technology, Rep. no. KHR7.

Vanoni, V.A., 1975 (ed.) *Sedimentation engineering*, Amer. Soc. Civil Engineers, New York.

Watson, C.C, Schumm, S.A. and Harvey, M.D., 1984. Neotectonic effects on river pattern. In C. Elliot (ed.), *River meandering*, Am. Soc. Civil Engineers, New York, 55-66.

Weir, G.J., 1983. One-dimensional bed wave movement in lowland rivers, *Water Resources Research* 19, 627-631.

White, W.R. and Day, T.J., 1982. Transport of graded gravel-bed material. In R.D. Hey, J.C. Bathurst and C.R. Thorne (eds.), *Gravel-bed rivers*, J. Wiley and Sons, 181-223.

Whittaker, J.C. and Jaeggi, M.N.R., 1982. Origin of step-pool systems in mountain streams, *Amer. Soc. Civil Engineers, J. Hydraulics Div.* 108, no. HY6, 758-722.

Williams, G.P. and Wolman, M.G., 1984. Downstream effects of dams on alluvial rivers, *US Geol. Surv. Prof. Paper*, 1286.

Wolman, M.G. and Brush, L.M., 1961. Factors controlling the size and shape of stream channels in coarse noncohesive sands, *US Geol. Surv. Prof. Paper 282G*, 183-210.

Yalin, M.S and Kanahan, 1979. Inception of sediment transport, *J. Hydraulics Div., Amer. Soc. Civil Engineers* 105, no. HY11, 1433-1443.

Yamaoka, I. and Hasegawa, K., 1984. Effects of bends and alternating bars on meander evolution. In C. Elliot (ed.), *River meandering*, Amer. Soc. Civil Engineers.

Yen, C.L., 1970. Bed topography effect on flow in a meander, *J. Hydraulics Div., Amer. Soc. Civil Engineers*, 97 (HY2), 303-321.

Ying, T., 1983. Intermittent surges of gravel transport in rivers, *Proc. 2nd international symp. on river sedimentation*, Water Resources and Electric Power Press, China, 368-377.

Zhengying, Q., 1983. The problems of river control of China, *Proc. 2nd international symp. on river sedimentation*, Water Resources and Electric Power Press, China, 8-33.

Index